Japan's Ocean Borderlands

Desert islands are the focus of intense geopolitical tensions in East Asia today, but they are also sites of nature conservation. In this global environmental history, Paul Kreitman shows how the politics of conservation have entangled with the politics of sovereignty since the emergence of the modern Japanese state in the mid-nineteenth century. Using case studies ranging from Hawai'i to the Bonin Islands to the Senkaku (Ch: Diaoyu) Isles to the South China Sea, he explores how bird islands on the distant margins of the Japanese archipelago and beyond transformed from sites of resource extraction to outposts of empire and from wartime battlegrounds to nature reserves. This study examines how interactions between birds, bird products, bureaucrats, speculators, sailors, soldiers, scientists and conservationists shaped ongoing claims to sovereignty over oceanic spaces. It considers what the history of desert islands shows us about imperial and post-imperial power, the web of political, economic and ecological connections between islands and oceans, and about the relationship between sovereignty, territory and environment in the modern world.

PAUL KREITMAN teaches modern Japanese history at Columbia University.

Cambridge Oceanic Histories

Edited by

David Armitage
Alison Bashford
Sujit Sivasundaram

Across the world, historians have taken an oceanic turn. New maritime histories offer fresh approaches to the study of global regions, and to long-distance and long-term connections. Cambridge Oceanic Histories includes studies across whole oceans (the Pacific, the Indian, the Atlantic) and particular seas (among them, the Mediterranean, the Caribbean, the North Sea, the Black Sea). The series is global in geography, ecumenical in historical method, and wide in temporal coverage, intended as a key repository for the most innovative transnational and world histories over the longue durée. It brings maritime history into productive conversation with other strands of historical research, including environmental history, legal history, intellectual history, labour history, cultural history, economic history and the history of science and technology. The editors invite studies that analyse the human and natural history of the world's oceans and seas from anywhere on the globe and from any and all historical periods.

Japan's Ocean Borderlands

Nature and Sovereignty

PAUL KREITMAN
Columbia University

CAMBRIDGE
UNIVERSITY PRESS

CAMBRIDGE
UNIVERSITY PRESS

Shaftesbury Road, Cambridge CB2 8EA, United Kingdom

One Liberty Plaza, 20th Floor, New York, NY 10006, USA

477 Williamstown Road, Port Melbourne, VIC 3207, Australia

314–321, 3rd Floor, Plot 3, Splendor Forum, Jasola District Centre, New Delhi – 110025, India

103 Penang Road, #05–06/07, Visioncrest Commercial, Singapore 238467

Cambridge University Press is part of Cambridge University Press & Assessment, a department of the University of Cambridge.

We share the University's mission to contribute to society through the pursuit of education, learning and research at the highest international levels of excellence.

www.cambridge.org
Information on this title: www.cambridge.org/9781108747462

DOI: 10.1017/9781108779241

First published 2023
First paperback edition 2025

A catalogue record for this publication is available from the British Library

Library of Congress Cataloging-in-Publication data
Names: Kreitman, Paul, 1983– author.
Title: Japan's ocean borderlands : nature and sovereignty / Paul Kreitman.
Description: Cambridge ; New York, NY : Cambridge University Press, 2023. | Series: Cambridge oceanic histories | Includes bibliographical references and index.
Identifiers: LCCN 2022055070 | ISBN 9781108489706 (hardback) | ISBN 9781108747462 (paperback) | ISBN 9781108779241 (ebook)
Subjects: LCSH: Japan – Environmental conditions. | Nature conservation – Political aspects – Japan. | Birds – Japan. | Island ecology – Japan. | Islands – Japan – History. | Borderlands – Japan – History. | Japan – Colonies – Oceania. | Sovereignty. | Imperialism.
Classification: LCC GE160.J3 K74 2023 | DDC 363.7/05610952–dc23/eng20230415
LC record available at https://lccn.loc.gov/2022055070

ISBN 978-1-108-48970-6 Hardback
ISBN 978-1-108-74746-2 Paperback

Contents

Maps

Figures

Acknowledgements

This book would not be possible without financial, intellectual and emotional support from a great many institutions and individuals. I began this project as a doctoral student at Princeton University and completed it at Columbia University. Along the way I was supported by research fellowships from the Japan Foundation, the Institute for Historical Research and the Japan Society for the Promotion of Science, and by a teaching fellowship at SOAS University of London. I am keenly aware that few other scholars have had access to the resources that enable this kind of multi-archival research across several continents, and I am grateful for the privilege.

As for individuals, it would surely be impossible to name every person who has helped guide this book to completion, amongst whom I must count not only teachers, colleagues and friends but also librarians, archivists and the support staff who keep the cogs of academia turning. Sponsors at several other universities have also invited me to present my research findings at conferences and lectures, providing opportunities for valuable feedback. Nevertheless, and in the sure knowledge I have overlooked someone crucial, I would like to single out the following people for special mention.

At Princeton I benefitted from a wise and patient adviser in the form of Shel Garon, but also indispensable guidance from David Howell, Emmanuel Kreike, Federico Marcon, Stephen Kotkin, Janet Chen and Su Naquin. I was also blessed by an unusually supportive graduate student cohort, among whom honourable mention goes to Erin Brightwell, Kjell Ericson, Yulia Frumer, Scott Moskowitz, Seiji Shirane, Wayne Soon, Megan Steffen, Jolyon Thomas, Kimberly Thomas and Ray Thornton. Beyond Princeton, Angus Lockyer and Taka Yamamoto provided crucial feedback and companionship during an interlude spent writing up the dissertation in London.

Among my colleagues at Columbia I would like to especially thank my academic mentors, David Lurie and Eugenia Lean, and also Greg

Pflugfelder, Haruo Shirane, Tomi Suzuki, Michael Como, Max Moerman, Carol Gluck, Takako Hikotani, Lien-Hang Nguyen, Matti Zelin, Hannah Farber, Karl Jacoby, Adam Tooze, Susan Pedersen, Rebecca Kobrin, Ria Koopmans-Debruijn, Sachie Noguchi and Laura Schlein. Plus of course my students, who have been a joy to teach even during the darkest days of COVID-19. Special thanks go to Peter Moody, Jenn Strange and Juzhi Zheng for their help with supplementary research. I must also thank Ross Yelsey at the Weatherhead East Asia Institute, who provided invaluable help crafting the book proposal.

In Japan, Yasutomi Ayumu kindly hosted me as a visiting researcher at the University of Tokyo's Institute for Advanced Studies on Asia. Fukao Yōko did likewise at Osaka University several years later, not only guiding me through the complexities of the JSPS application process but providing myself and my family sanctuary during the worst of the pandemic. Sakaguchi Makoto and Ikeda Ryō were good enough to invite me to join the Fertilizer History Research Group and the Informal Empire Research Project respectively. Hasegawa Hiroshi and Yokohata Yasushi both generously shared their time and expertise with me, a non-scientist, as did the staff of the Yamashina Institute for Ornithology. Kuniyoshi Masafuru shared valuable documentary and visual material related to the history of the Senkaku Isles. And Takara Tetsumi enlightened me on his father's human side and the taste of dugong meat.

Countless others have also provided support, intellectual stimulation and friendship over the years, including (but not limited to) Araragi Shinzō, Dani Botsman, Ted Boyle, Sakura Christmas, Fabian Drixler, David Fedman, Gal Gvili, Toshi Higuchi, Nadin Heé, Stefan Hübner, Mariko Iijima, Tatiana Linkhoeva, Aike Rots, Jonas Rüegg, Jordan Sand, Sujit Sivasundaram, Tehila Sasson, Ariel Stilerman, Bill Tsutsui, plus of course my late undergraduate adviser, Ann Waswo. Ian Jared Miller, Alexis Dudden, Greg Cushman and Martin Dusinberre all provided crucial feedback on draft versions of this manuscript at various stages. And Lucy Rhymer and Laheba Alam at Cambridge University Press and Reshma Xavier at Integra have been unusually patient and understanding editors throughout the long slog to completion.

Finally, renewed thanks and love to my family – to Jojo, Monty, Tricia, Roger, Oliver, Eliza, Arthur and Richard, but also to the Iwase family in Tokyo and among them Noriko, who passed long before her time was due. I dedicate this book to her memory.

Naming Conventions

Many of the islands discussed in this book have been known by various names in different languages, and the choice of one name over others risks being construed as politically loaded. So I should say at the outset that this book does not seek to adjudicate territorial sovereignty claims. Instead, when choosing which names to use for islands or island groups I have generally tried to prioritise clarity over politics where possible.

This means I have not been entirely systematic. In some cases I have opted for names that sound more familiar to readers in English: so Iwo Jima over Iōtō, Marcus Island over Minami-torishima, Rasa Island over Okidaitōjima, Sakhalin over Karafuto, Guam over Guåhan, Sea of Japan over East Sea. In other cases I have echoed the names used by my sources, particularly when I am interested in recovering the particular historical perspectives of those sources. In Chapter 7, for instance, which is told largely from Okinawan and Japanese perspectives, I refer to the Senkaku Islands rather than the Diaoyu-tai, Diaoyu-dao or Pinnacle Islands. This by no means connotes support for the Japanese government's territorial claim to the islands. To avoid confusion I have generally tried to refer to islands by the same names throughout, even though their 'official' names may have changed multiple times. So I have stuck with Bonin over Ogasawara Islands even when discussing their time under Japanese rule.

Though sensitive to the cause of Indigenous rights, I have also avoided using anachronistic names where possible. So in referring to the Indigenous lands of the Ainu I use the settler colonial name Hokkaidō over the more recent coinage Ainu Mosir. Conversely in Chapter 7, I refer to Okinawans rather than Ryūkyūans because, notwithstanding the historical existence of an independent Ryūkyū Kingdom, I detected little evidence of sustained Ryūkyūan nationalist sentiment during the period of US colonial rule over the islands. Finally, when writing about the Northwest Hawaiian Islands I have used

Laysan, Lisianski and Midway rather than the Hawaiian-language names proposed by Kekuewa Kikiloi.[1] Though the names recovered by Kikiloi effectively articulate traditional Indigenous claims to the islands (and may well have been used prior to the nineteenth century), by the early twentieth century they appear to have fallen out of use even amongst Native Hawaiians (Kānaka Maoli or Kānaka ʻŌiwi. I have, however, listed Indigenous names for Hawaiian and other Pacific islands where they first appear in the text.

Japanese and Chinese names are written in the East Asian fashion – surname before given name – except when citing works published in English. I have added macrons to Japanese vowels where appropriate, but not for places such as Tokyo or Osaka that are well known to English-language readers. Chinese words and names are mainly written in pinyin. I have included ʻokina and kahakō in Hawaiian terms and names unless quoting primary sources that do not include them.

[1] Kekuewa Kikiloi, 'Rebirth of an Archipelago: Sustaining a Hawaiian Cultural Identity for People and Homeland', *Hūlili* 6 (2010): 73–115.

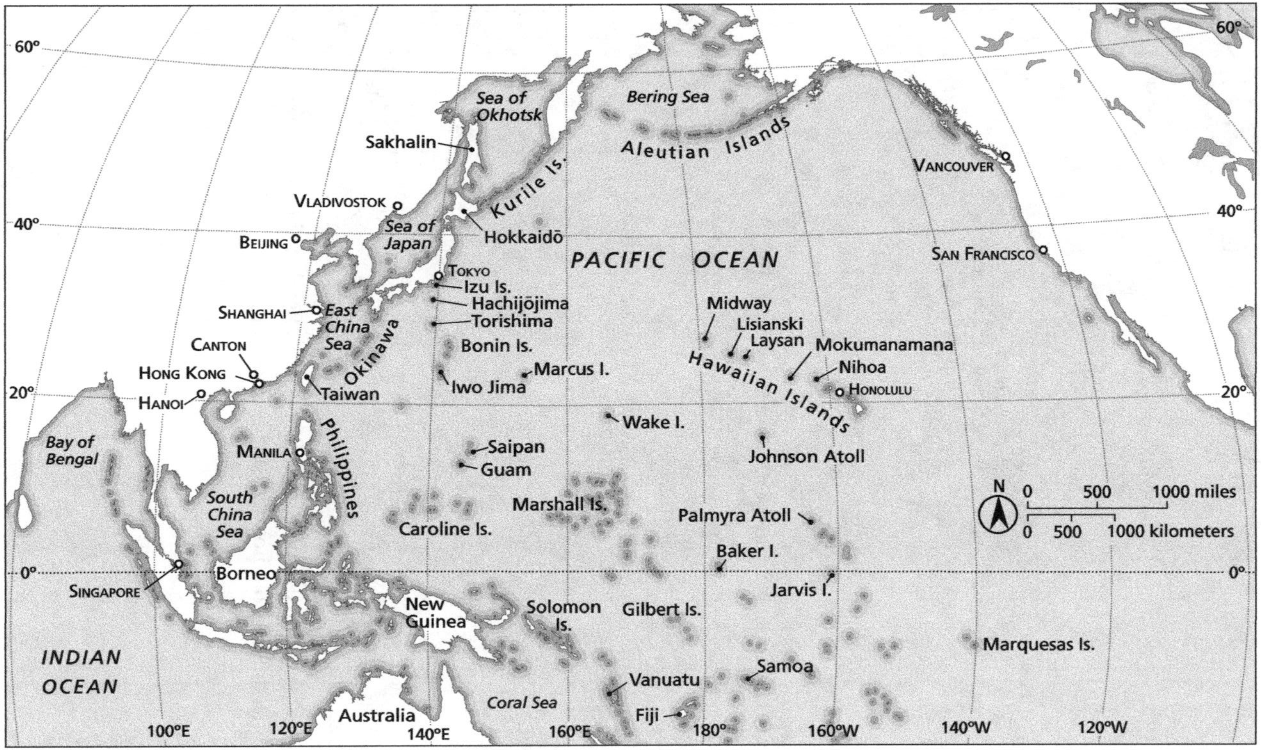

I The North Pacific Ocean

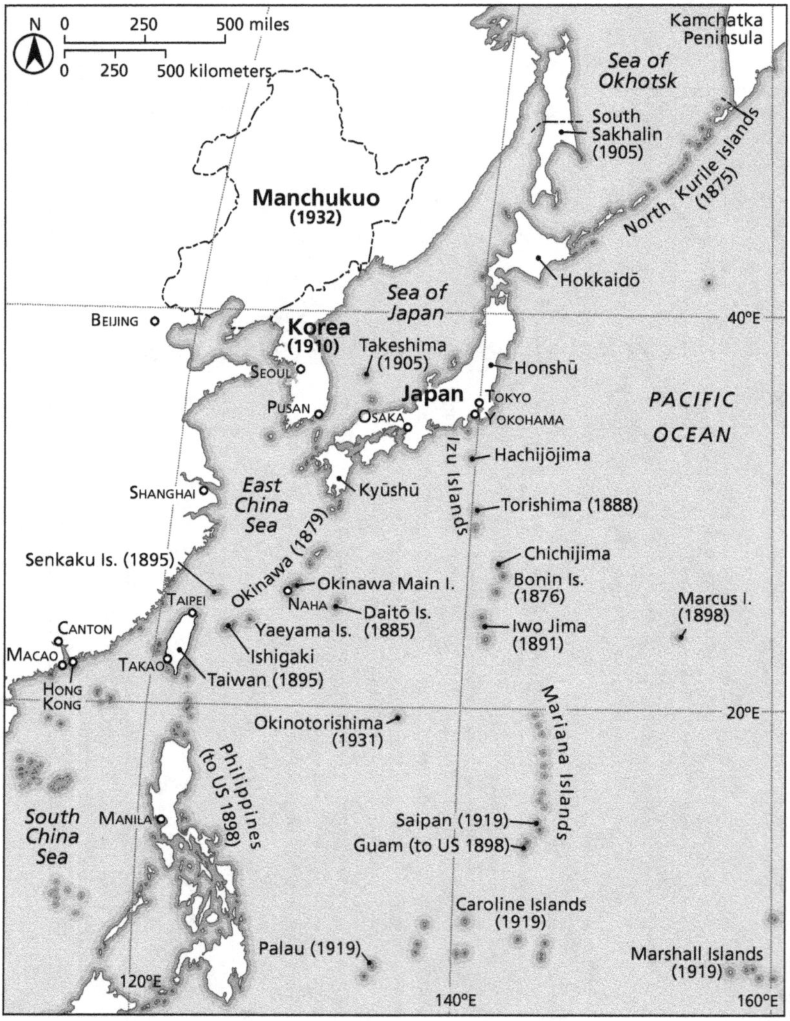

II Japanese colonial expansion (1875–1933). Bracketed numbers refer to year of formal annexation, not date of effective control

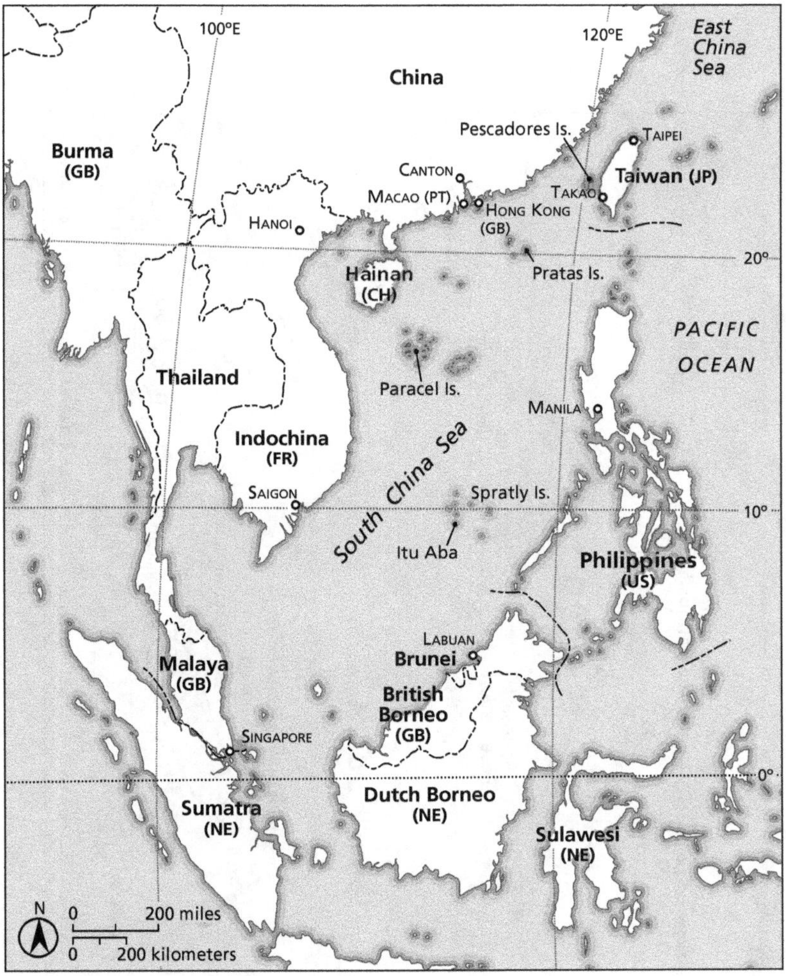

III The South China Sea (c.1930)

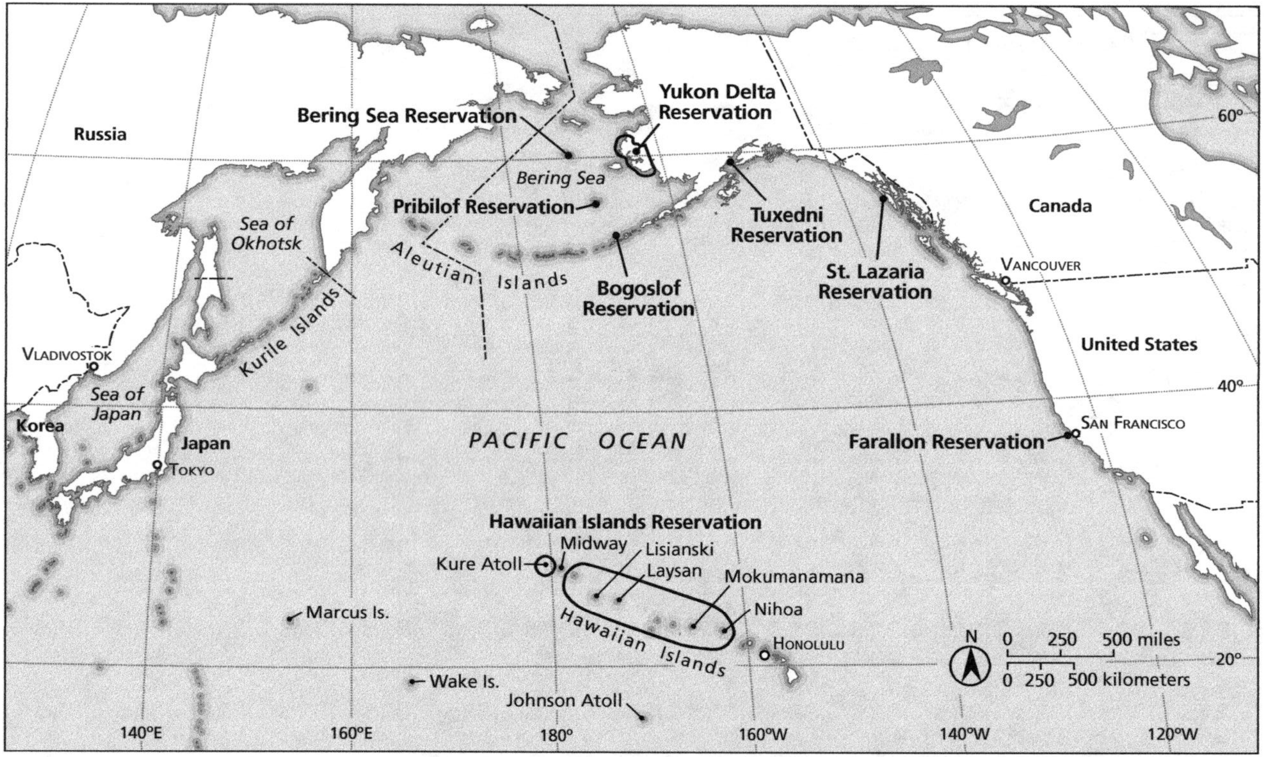

IV Bird reservations established by executive order of Theodore Roosevelt (February–March 1909)

Introduction

The idea for this book hatched nearly a decade ago, during an aimless afternoon spent browsing in Tokyo's bookselling district. I stumbled across a poster showing a flock of huge birds wheeling above a lush tropical island (Figure 0.1). An advertisement for guano fertiliser mined on a mysterious isle called 'Rasa' a century ago, it would once have hung on the wall of a village wholesaler or agricultural co-op. Buy our product, it seemed to promise, and channel the energy of the ocean, routed through the bowels of these birds, into your own crops.

At the time I had never heard of Rasa, but a quick google search brought up a satellite image of a barren, foam-ringed chunk of rock in the middle of the North Pacific. Devoid of vegetation, scarred by years of open-face mining, the island's present state is a far cry from the tropical idyll shown in the advertisement (Figure 0.2). Though technically administered as part of the Japanese prefecture of Okinawa, it is today uninhabited. The birds are long gone, too. In fact, the species depicted in the poster, Steller's albatross, was likely wiped out on the island even before the first guano miners arrived – its long wing-feathers plucked and sold to adorn the hats of women in Paris, London and New York.

Digging further, I discovered that Rasa was only one of a number of uninhabited islands that ring the outer perimeter of the Japanese body politic. The significance of these islands is entirely disproportionate to their tiny size. During Japan's colonial period they unfurled vast riches into the laps of the prospectors who harvested their resources for a quick profit, and much more modest incomes for their employees: the men and women who did the actual work of butchering seabirds and shovelling their faeces onto ships. The islands also generated reams of public discourse that fetishised them as symbols of colonial expansion, testimony to the pioneer spirit of the Japanese nation. Many went on to become key nodes in Japan's defence perimeter. One, Iwo Jima,

Figure 0.1 Advertisement for Rasa Island guano (n.d.)

became the site of one of the bloodiest battles in the Pacific War. Later still, even after their resources were depleted and strategic value spent, bird islands became symbols of nationalist pride and anchors of pelagic sovereignty under new regimes of international maritime law. They remain so to this day.

Intriguingly, some of these islands have also become sites for nature conservation. It was once thought that the activities of Japanese hunters drove Steller's albatross to extinction, but in the post-war era the bird has been resurrected by conservationists as a 'natural monument of the nation', symbolising Japan's commitment to peace, democracy and enlightened stewardship of the environment. It was, for example, one of the earliest species added to the International Union for the Conservation of Nature's global database

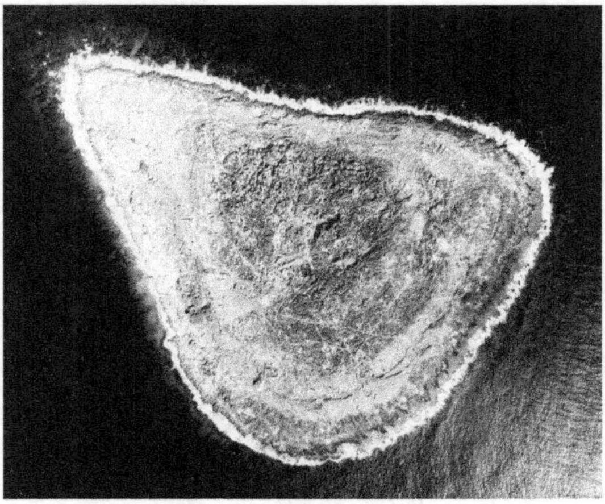

Figure 0.2 Aerial photograph of Rasa Island (1978)
© National Land Image Information, Ministry of Land, Infrastructure, Transport and Tourism

of threatened species. To this end, many of Japan's remote islands have been incorporated into national parks and United Nations Educational, Cultural and Scientific Organisation (UNESCO) world heritage sites. At the time of writing numerous projects are under way to repopulate them with rare seabirds, wipe out invasive species such as rats and goats, and restore them to something approximating their prior ecological state.

Yet these conservation efforts have a contentious history. The first albatross conservation campaigns were mounted at the turn of the twentieth century by American ornithologists, who sought to evict Japanese bird-hunters from recently annexed Hawaiian atolls at a time of heightened anxiety about trans-Pacific Asian migration. Japanese conservation campaigns, for their part, have become entangled with the politics of the US military presence in East Asia and with festering territorial disputes with Japan's neighbours. As recently as 2013, during a bitter confrontation with China over the sovereign status of the Senkaku (Ch. Diaoyu) Islands, a conservation charity linked to the Japanese royal family suggested the government send a landing party to search for a potential new subspecies of Steller's albatross. The suggestion triggered a fiery response from Chinese state media, which denounced a Japanese plot to 'snatch our islands using the pretext of environmental protection'.

This book sets out to answer the flood of questions the Rasa Island guano poster raised for me. What chain of events led people to set foot on such remote spots in the first place? How did they go about claiming the birds and the land on which they nested? What kind of human settlements once existed on the islands, and what happened after they were abandoned? What does the history of bird islands say about Japanese imperial and post-imperial power, and about the web of political, economic and ecological connections between insular and oceanic space? What changed between the age of bird butchery and the age of bird protection – and what has stayed the same? And what does all this say about the relationship between sovereignty, territory and environment in the modern world? Did China's accusation of a conservationist conspiracy to 'snatch' bird islands not, perhaps, have a grain of truth to it?

In addressing these questions, this study sits at the intersection of three overlapping fields of historical enquiry: Japanese colonialism, Pacific Ocean studies and environmental history. Because of this, I have aimed to write it in such a way that readers can grasp the narrative and core arguments without deep background knowledge of any of these fields. In the pages that follow I will set out how the study contributes to each field in turn, beginning with the relatively narrow field of Japanese history and moving outwards to encompass the Pacific world more generally, before finally addressing the broader subdiscipline of environmental history. It is always a challenge to write a book that speaks to multiple subfields in this manner, but I hope that the pages herein contain something for everyone.

Shards of Empire?

Bird islands occupy a peculiar space within the Japanese body politic. They were first colonised during a century of expansion that began in the 1850s and ended abruptly in 1945: in this sense they are what Alexis Dudden has called 'shard-like remnants of the nation's once massive empire'.[1] It is now well known that Japan once had an overseas empire of colonies and puppet states. Key possessions included Taiwan (annexed in 1895), Southern Sakhalin (1905), Korea (1910), Micronesia (1914), Manchukuo (1931–1932), plus the brief Greater East Asia Co-Prosperity Sphere (1940–1945), which at its zenith

[1] Alexis Dudden, 'Japan's Island Problem', *Dissent* (Fall 2014): 12.

stretched from Mongolia to Burma to Indonesia to the Aleutian Islands. The history of Japan's bird islands cannot be grasped without reference to this broader territorial empire. It is hardly a coincidence, for example, that Japan chose to annex the Senkaku Islands in the middle of a war with neighbouring Qing China. And many of the same people who helped to colonise bird islands such as Torishima and Marcus Island were also active elsewhere in the Japanese Empire, particularly in Micronesia.

More recent scholarship, however, has sought to move beyond empire studies in the traditional sense in a number of ways. Some historians have chipped away at the conventional distinction between colony and metropole by drawing attention to 'internal colonies' such as Hokkaidō and Okinawa, which remain part of the Japanese body politic to this day.[2] Others have drawn attention to what I here call 'extraterritorial empire', showing how gunboat diplomacy secured consular protection for Japanese imperialists even when they strayed well beyond the limits of Japanese-administered territory.[3] At the same time, other scholars have drawn connections between Japanese colonial expansion and the migration of Japanese subjects across the Pacific, suturing together the once divided fields of East Asian history and Asian-American studies.[4] The most ambitious work has followed this line of thinking to its logical conclusion by

2 For a discussion of Japan's internal colonies see Takaesu Masaya, *Kindai Nihon to shokuminchi to 'tōsho'* (Yumani Shobō, 2009), viii; Michele Mason and Helen Lee, *Reading Colonial Japan: Text, Context, and Critique* (Stanford University Press, 2012), 18–20. The contested identity politics of Okinawa are discussed in Chapter 7.

3 Peter Duus, Ramon H. Myers, Mark Peattie (eds), *The Japanese Informal Empire in China, 1895–1937* (Princeton: Princeton University Press, 1989); Erik Esselstrom, *Crossing Empire's Edge: Foreign Ministry Police and Japanese Expansionism in Northeast Asia* (Honolulu: University of Hawai'i Press, 2008); Seiji Shirane, *Imperial Gateway: Colonial Taiwan and Japan's Expansion in South China and Southeast Asia, 1895–1945* (Ithaca: Cornell University Press, 2022).

4 Shiode Hiroyuki, *Ekkyosha no Seijishi: Ajia Taiheiyō ni okeru Nihonjin no imin to shokumin* (Nagoya: Nagoya Daigaku Shuppankai, 2015); Eiichiro Azuma, *In Search of Our Frontier: Japanese America and Settler Colonialism in the Construction of Japan's Borderless Empire* (Oakland: University of California Press, 2019); Sidney Xu Lu, *The Making of Japanese Settler Colonialism: Malthusianism and Trans-Pacific Migration, 1868–1961* (Cambridge: Cambridge University Press, 2019); Martin Dusinberre & Mariko Iijima, 'Transplantation and Imperial Practice in Japan's Pacific', *in Historische Anthropologie* 27 no. 3 (December 2019): 325–32.

historicising the production of territorial space itself.[5] The result has been to destabilise conventional distinctions between nations and empires and instead highlight the emergence of centrally administered, territorially demarcated states as a profound historic rupture in its own right.[6]

Pacific Frontier or Ocean Borderland?

Given all these recent innovations, it becomes apparent that describing Japanese activities on bird islands as 'colonial' only gets us so far. How then should we locate bird islands within the sprawling expanse of the Japanese Empire? What should we make, for instance, of the fact that many were governed not as colonial appendages but as part of the Japanese metropole itself? Despite lying twelve hundred kilometres out to sea, for instance, the island of Iwo Jima is technically part of the city of Tokyo. From the outset I knew that explaining the peculiarity of bird islands within the Japanese body politic would require a particular spatial analytic.

One option that presented itself was the concept of the frontier. In some ways the term is an attractive one. For one thing, it has long been used to describe the expansion of human settlements into wild and/or uninhabited space. And while this framing has often served to erase or marginalise the existence of Indigenous people, in fact bird islands were unique within Japan's territorial empire in that they mostly *did* lack a history of human habitation. One exception to this rule, the Bonin Islands (J: Ogasawara Shotō), is discussed in Chapter 1, but in general the colonisation of bird islands was quite distinct from imperial rule over populated territories such as Taiwan, Micronesia and Hawai'i. If the latter cases fit recent understandings of colonialism as the 'occupation of an already inhabited island or locale by another population seeking to displace or dominate the

5 Tessa Morris-Suzuki, *Borderline Japan: Frontier Controls, Foreigners and the Nation in the Post-War Era*. (Cambridge: Cambridge University Press, 2010); Pär Kristoffer Cassel, *Grounds of Judgment: Extraterritoriality and Imperial Power in Nineteenth-Century China and Japan* (Oxford: Oxford University Press, 2012); David Ambaras, *Japan's Imperial Underworlds: Intimate Encounters at the Borders of Empire* (Cambridge: Cambridge University Press, 2018).

6 Charles Maier, 'Leviathan 2.0', in *A World Connecting, 1870–1945*, ed. Emily S. Rosenberg (Cambridge, MA: Harvard University Press, 2012). On the pre-territorial Japanese 'geography of civilization' see David Howell, *Geographies of Identity in Nineteenth-Century Japan* (Oakland: University of California Press, 2005): Ch. 6.

original inhabitants', then bird islands were closer to the older, Latin-derived sense of the word: *colere,* meaning simply 'to cultivate, or inhabit'.[7]

Other scholars have deployed the term 'frontier' in a more narrowly economic sense, for instance to describe the interplay between capitalist and non-capitalist space. Indeed, in some ways this book charts the emergence and collapse of a 'commodity frontier zone' predicated on extractive economic activity.[8] But I am at pains to emphasise that this was not a unidirectional process. Capitalist expansion was frequently undermined by ecological 'frictions': resource exhaustion due to unsustainable extraction, but also fluctuations in weather and climate, tidal erosion and volcanic activity. The economic viability of oceanic frontier zones was also hostage to 'social frictions'. Working conditions onshore bird islands were often grim, and guano mining operations in particular plagued by frequent labour disputes. The profitability of bird island settlements was also hostage to whiplashing commodity prices that were in turn driven by a host of other factors: boom and bust cycles in the global economy, the availability of substitute goods (kerosene for whale oil, chemical fertiliser for guano) and the sheer vagaries of consumer fashion in the case of plumage.

Finally, as scholars such as Eichiro Azuma and Sidney Xu Lu have shown, many self-styled colonial pioneers did indeed employ frontier-inflected terms like *hatten* ('development'), *shokuminchi* ('colony'), *bōchō* ('expansion') and *kaitaku* ('colonise/develop') – both to fashion their own sense of identity and to whip up support for their efforts.[9] The bird-hunters and guano miners who led the effort to colonise bird islands were equally fond of such language. In this sense the frontier as an imagined construct, conjured into being as part of an ambient 'culture of imperialism', was a very real thing.[10]

7 Glen Petersen, 'Indigenous Island Empires: Yap and Tonga Considered', *The Journal of Pacific History* 35 no.1 (June 2000): 6.

8 Sven Beckert, Ulbe Bosma, Mindi Schneider and Eric Vanhaute, 'Commodity Frontiers and the Transformation of the Global Countryside: A Research Agenda', *Journal of Global History* 16 no. 3 (November 2021): 435–50.

9 Azuma, *In Search of Our Frontier,* 14–15; Lu, *The Making of Japanese Settler Colonialism,* 31–4.

10 Louise Young, *Japan's Total Empire: Manchuria and the Culture of Wartime Imperialism* (Oakland: University of California Press, 1999); Aaron Skabelund, *Empire of Dogs: Canines, Japan, and the Making of the Modern Imperial World* (Ithaca: Cornell University Press, 2011).

But the very fact that empire-builders (and wannabe empire-builders) were so fond of frontier-style rhetoric should caution us against applying it uncritically ourselves. A particular hazard of the term 'frontier' is that it risks treating pre-existing civilisations (or nations or empires if you prefer) as a priori units of analysis.[11] This tendency is pervasive throughout much historical writing: many studies that do not explicitly use the term 'frontier' nevertheless echo its premise by referring simply to Japanese 'expansion' without specifying exactly what they mean by 'Japan'. One influential study of US–Japan relations was entitled simply 'The Clash', giving the impression of two ontologically stable geopolitical entities facing off against each other from the moment of their first encounter until the trade war of the 1980s.[12]

Yet colonial, imperial, economic and territorial boundaries did not move in lockstep, so in this study I prefer to use the term 'borderland'. If frontiers are bilateral, unidirectional and teleological, borderlands are 'ambiguous and often unstable realms where boundaries are also crossroads, peripheries are also central places, homelands are also passing-through places, and the end points of empire are also forks in the road'.[13] A borderland perspective permits what Tessa Morris-Suzuki has called an 'anti-area studies' approach: 'by placing oneself not within the boundaries of traditionally defined world regions ... but precisely *on* the boundaries ... the focus of research then becomes the shifting and fluid nature of these boundaries themselves'.[14] The sociologist Ishihara Shun likewise argues that the oceanic Northwest Pacific is an ideal site from which to conduct a 'fixed point observation of modernity' – by which he means the interlinked processes of globalisation, state-making and colonialism.[15] It allows us, in other words, to examine processes of

[11] Richard White, Patricia Nelson Limerick and James R. Grossman, *The Frontier in American Culture* (Oakland: University of California Press, 1994).

[12] Walter Lafeber, *The Clash: A History of US–Japan Relations* (New York: W.W. Norton & Co, 1997).

[13] Pekka Hämäläinen and Samuel Truett, 'On Borderlands', *The Journal of American History* 98 no. 2 (2011): 338; Jeremy Adelman and Stephen Aron, 'From Borderlands to Borders: Empires, Nation-States, and the Peoples in Between in North American History', *American Historical Review* 104 no. 3 (June 1999): 816.

[14] Tessa Morris-Suzuki, *On the Frontiers of History: Rethinking East Asian Borders* (Canberra: ANU Press, 2020), 22.

[15] Ishihara Shun, *'Guntō' no rekishi shakaigaku: Ogasawara Shotō, Iōtō, Nihon, America, soshite Taiheiyō sekai* (Tokyo: Kōbundō, 2013), 21–2.

modern state formation without treating the state itself as a pre-existing unit of analysis.

Imagining the Pacific as an oceanic borderland is fruitful for other reasons. For one thing it places the traditional bilateral story of US–Japan relations within a broader international context, highlighting not only the importance of Britain as a global hegemon in the nineteenth century but also the role of smaller regional players such as the Kingdom of Hawai'i, not to mention the cast of beachcombers and Pacific Islanders who drifted around the region during the period.[16] Conversely it shows the role of the United States in shaping post-war territorial disputes between ostensibly independent East Asian governments. Throughout, it helps highlight the crucial role of exterritoriality in shaping contests over terrain, people and resources. To put it another way, it reminds us that the territorialisation of sovereignty is a recent and still incomplete phenomenon, and that for much of its history the Pacific has been what Ziad Fahmy has called a 'jurisdictional borderland'.[17]

A borderland approach also reveals the dialectical processes that helped to propel Pacific state-making. Not only was Japan far from alone in practising a form of 'mimetic imperialism' but Japanese activity in the Pacific also spurred rival actors to develop new techniques of asserting sovereignty over bird islands.[18] In fact, although the islands discussed in this book arc loosely around the periphery of the contemporary Japanese state, the perspective is by no means uniquely Japan-centric. Some islands, such as the Bonin Islands, are today Japanese territory but were originally settled by Pacific Islanders and white 'beachcombers'. Others, such as the Northwest Hawaiian Islands, have never been Japanese territory in any formal diplomatic sense but were once sites of economic activity by Japanese nationals. Some, such as the Spratly Islands in the South China Sea, were annexed by Japan but relinquished after World War II. And others still, such as the

16 Lorenz Gonschor, *A Power in the World: The Hawaiian Kingdom in Oceania* (Honolulu: University of Hawai'i Press, 2019); Petersen, 'Indigenous Island Empires'.

17 Ziad Fahmy, 'Jurisdictional Borderlands: Extraterritoriality and "Legal Chameleons" in Precolonial Alexandria, 1840–1870', *Comparative Studies in History and Society*, 55 no. 2 (2013): 305–29.

18 Robert Eskildsen, 'Of Civilization and Savages: The Mimetic Imperialism of Japan's 1874 Expedition to Taiwan', *The American Historical Review* 107 no. 2 (2002): 388–418.

Senkaku Islands and Takeshima (Korean: Dokdo), remain fiercely contested by neighbouring states to this day. The view from bird islands, in other words, offers a fresh perspective not only on Japan but also on East Asia, the Pacific and still more distant shores.

Environment and Empire

A borderland approach to the Pacific offers another advantage, in that it helps to highlight the agency of the oceanic environment itself. Bird islands are the products of long-term environmental processes such as seabed volcanic activity, circulatory currents, tidal erosion and sedimentation, migration by plants and animals, trophic cascades and coral growth. Oceanic ecologies sustained the webs of life that fuelled extractive industries such as whaling, bird-hunting and guano mining, and the ocean's liquid 'low friction' terrain enabled a degree of maritime 'borderless mobility' that is impossible on land.[19]

Yet at the same time, the oceanic environment frequently frustrated attempts at permanent human settlement. The sheer inaccessibility of bird islands – which made them ideal nesting sites in the first place – acted as a barrier to humans much as it did to other land-borne predators. Their scarcity of potable water meant they could support only limited human populations without recourse to expensive desalination infrastructure. (Seabirds are not troubled by the lack of freshwater, for their bills are connected to salt glands that allow them to drink seawater.)[20] Moreover, bird islands' tiny surface areas generally contained negligible amounts of arable land. Bird hunters and guano miners in particular suffered from the scourge of beri-beri, a diseased cause by a vitamin-deficient diet.[21] Seasonality posed another challenge to human settlement. The seabirds on whom plumage hunters depended nested on the islands only at certain times of the year, and

[19] For 'low friction terrain' see James C. Scott, *The Art of Not Being Governed: An Anarchist History of Upland Southeast Asia* (New Haven: Yale University Press, 2009), 330 n8. For 'borderless mobility' see Ishihara Shun, *Kindai Nihon to Ogasawara-tō: idōmin no shimajima to teikoku* (Tokyo: Heibonsha, 2007), Ch.3.

[20] William A. Buttemer, J. Eduardo P. W.Bicudo, Mark A. Chappell, James T. Pearson and Claus Bech, *Ecological and Environmental Physiology of Birds* (Oxford: Oxford University Press, 2010), 171.

[21] Noritaka Tsunetō, *Discovery and Exploitation of Rasa Island* (Tokyo: Rasa Rinkō Kabushiki Gaisha, 1920) 35; MOFA 3.5.2.8: Rear Admiral Evans to Ambassador, Tokyo (6 January 1903).

in summer and early autumn the islands and the seas around them were ravaged by typhoons that swept away dwellings and sank the ships that provided the lifeline to the fragile colonies. For all these reasons the rhetoric of the colonial pioneer often foundered on the rocks of reality, sometimes quite literally.

Even if all these obstacles were somehow overcome, there remained the fact that most bird island colonies were fundamentally predicated on unsustainable extractive industries. The prospectors who petitioned to lease or buy the islands usually declared their intention to establish permanent settlements based on agriculture or fishing, and some did indeed attempt to follow through on these promises by investing in cattle ranching or sugar cane plantation. But the truth was that the capitalists who financed the 'development' of bird islands were primarily interested in extracting their resources to harvest a quick profit. After their birds had been slaughtered and their guano deposits tapped out, the islands themselves generally made for unappetising economic propositions.

This brings me to the book's third intervention, on the relationship between imperialism, resource extraction and sustainability. There is already a rich body of work mapping the environmental history of resource extraction. Many readers will already be familiar with Gregory Cushman's magisterial book on *Guano and the Opening of the Pacific World* – which frames the scramble for bird islands as a form of 'neo-ecological neo-imperialism' by which European and Neo-European societies wreaked global environmental havoc in order to control and extract the resources necessary to fuel unsustainable growth.[22] And Toshi Higuchi has suggested that Japan's historical reliance on imported fertiliser biomass such as Okhotsk herring and Manchurian soybeans can be understood as a form of 'organic imperialism'.[23] In a slightly different vein, Brett Walker has also framed Japanese colonial expansion in Hokkaidō as a form of ecological imperialism.[24]

22 Gregory T. Cushman, *Guano and the Opening of the Pacific World: A Global Ecological History* (Cambridge University Press, 2013), 19.

23 Toshihiro Higuchi, 'Japan as an Organic Empire: Commercial Fertilizers, Nitrogen Supply, and Japan's Core-Peripheral Relationship', in *Environment and Society in the Japanese Islands: From Prehistory to the Present*, ed. Philip Brown and Bruce Batten (Portland: University of Oregon Press, 2015).

24 Brett L. Walker, *The Lost Wolves of Japan* (Seattle: University of Washington Press, 2005); Brett L. Walker, *The Conquest of Ainu Lands: Ecology and Culture in Japanese Expansion, 1590–1800* (Oakland: University of California Press, 2006).

The recent oceanic turn in Japanese history has likewise been inflected with a strong environmental dimension, with William Tsutsui suggesting that Japan was (and perhaps still is) a 'pelagic empire' based on 'domination and exploitation of offshore fisheries in the surrounding seas and, eventually, much of the western Pacific Ocean'.[25] Hiraoka Akitoshi, to whose research the current work is particularly indebted, has argued that Japanese imperial expansion in the Pacific was driven in part by the unsustainable extraction of bird island resources such as plumage and guano.[26] These are only a few of the many scholars who have stressed the connections between capitalism, imperial expansion and unsustainable economic activity.

This study is deeply indebted to this earlier body of research. But the more I dug into the history of bird islands, the more I saw an opportunity to tell another side of the story as well. My hunch was that if you were to analyse efforts to save birds in the same way that you analysed efforts to kill them, it would result in a much richer, knottier and morally ambivalent story. For one thing, empires are not always heedless of environmental destruction. Richard Grove's *Green Imperialism* shows how the experience of French and British empire-builders on Indian Ocean islands such as Mauritius and Réunion helped to shape concepts of ecology and sustainable resource use.[27] And David Fedman has stressed that, although Japanese control over the forests of the Korean peninsula 'grew in tandem with Japan's imperial sphere', colonial foresters were for reasons of sheer self-interest keenly interested in managing forests sustainably where possible.[28]

Fedman's research builds on the work of many other scholars who have stressed that nature conservation is, just as much as resource

[25] William Tsutsui, 'The Pelagic Empire: Reconsidering Japanese Expansion', in *Japan At Nature's Edge: The Environmental Context of a Global Power*, ed. Ian Jared Miller, Julia Adeney Thomas and Brett L. Walker (Honolulu: University of Hawai'i Press, 2013), 23.

[26] See in particular Hiraoka Akitoshi, *Ahōdori to 'teikoku' Nihon no kakudai: Nan'yō no shimajima e no shinshutsu kara shinryaku e* (Tokyo: Akashi Shoten, 2012), translated into English as Akitoshi Hiraoka, *Japanese Advance into the Pacific Ocean: The Albatross and the Great Bird Rush* (Singapore: Springer, 2018).

[27] Richard H. Grove, *Green Imperialism: Colonial Expansion, Tropical Island Edens and the Origins of Environmentalism, 1600–1860* (Cambridge University Press, 1996).

[28] David Fedman, *Seeds of Control: Japan's Empire of Forestry in Colonial Korea* (Seattle: University of Washington Press, 2020).

extraction, a political project. The Manchu rulers of Qing China, for instance, set aside vast swathes of their 'wild' homeland in Manchuria as a way of reinforcing ethnic distinctions and burnishing imperial authority.[29] Paradigmatically, William Cronon has shown how early nature conservationists were rich white men who sought to preserve 'wilderness' as a recreational space in which to balm their own anxieties about race, class and their own virility.[30] Achieving these ends often involved denying Indigenous populations access to game, firewood and grazing land. Ever since, national park management around the world has provoked fierce debates about how, by whom and for whom specific environments ought to be managed.

Nature and Sovereignty Conservation

We can see similar dynamics playing out within borderland space. Tellingly, the first recorded denunciations of Japanese bird-hunting came from American bird conservationists – who were also deeply worried about defending their own country's imperial presence in the Pacific. Indeed, debates on how to appropriately manage the environments of bird islands were, from the very start, intermingled with debates over territorial sovereignty. It is particularly striking that many state efforts to conserve the environments of bird islands got under way only after initial colonisation attempts had failed – after the narrative of 'frontier'-style unidirectional development had faltered.

This in turn made me want to know more about the afterlives of bird island colonies. What happened after the prospectors moved on? Even after they ceased to be sites of resource extraction many bird islands accrued new value. Some became freighted with immense strategic importance, and indeed became the sites of bitterly fought battles during the Pacific War. And recent developments in international maritime law have transformed them into anchors of pelagic sovereignty. Others have been seized on by 'national identity entrepreneurs' as a synecdoche for the nation itself.[31] As a result, territorial disputes

29 Jonathan Schlesinger, *A World Trimmed with Fur: Wild Things, Pristine Places, and the Natural Fringes of Qing* (Stanford: Stanford University Press, 2017).

30 William Cronon, 'The Trouble with Wilderness; or, Getting Back to the Wrong Nature', *Environmental History* 1 no. 1 (January 1996): 7–28.

31 Alexander Bukh, *These Islands Are Ours: The Social Construction of Territorial Disputes in Northeast Asia* (Stanford: Stanford University Press, 2020), 10.

over bird islands have continued long after they ceased to have any economic importance in their own right.

Efforts to conserve nature on abandoned bird islands occurred within this context, as part of a range of performative strategies aimed at projecting 'the effect of the state' onto contested territory.[32] The contrast with earlier types of performance is instructive. Earlier claimants had claimed sovereignty over bird islands through a variety of means. Some performances, such as rituals of naming, or notifications of annexation dispatched to foreign ministries, were primarily discursive in nature. But others also had a material component, ranging from the raising of flags and the erecting of stele to the installation of physical infrastructure such as wells, harbours and meteorological stations. We might call these latter strategies performances of colonisation, as they attempted to create the impression of settlement on the islands. This generally entailed attempting to maintain a human presence, ideally comprising men, women and children engaged in a non-extractive economic activity such as agriculture, and served by schools, medical facilities, government offices, graveyards and other amenities necessary to establish permanent habitation. Failing that, claimants often drew up elaborate plans suggesting that seasonal or intermittent extractive industries such as plumage gathering or guano mining constituted a beachhead for permanent colonisation.

If this first form of performance claimed to render remote islands habitable by butchering their bird populations, performances of conservation essentially inverted this strategy. Beginning in the early twentieth century, states and their allies began to use not environmental transformation but *conservation* as a means to assert sovereignty over territory. Conservation of nature in its pristine state, rather than the development of it, emerged as a 'political technology' for defending sovereignty.[33]

On the surface, the performance of nature conservation appeared to be a form of enlightened governance, an inversion of prior extractivist attitudes to bird islands. And there is no question that at least some of its advocates sincerely believed in the importance of bird conservation as an end in itself. But their activity, like those earlier performances,

[32] Timothy Mitchell, 'The Limits of the State: Beyond Statist Approaches and Their Critics', *American Political Science Review* 85 no. 01 (1991): 78.

[33] For a discussion of territory itself a political technology see Stuart Elden, *The Birth of Territory* (Chicago: University of Chicago Press, 2013), 315–20.

also served to perform sovereignty. In many cases it involved efforts to expel foreign actors from putatively national space. In others it connected to broader projects to bolster national sovereignty. In post-war Japan, for instance, Japanese ornithologists set about transforming the bird island of Izu-Torishima into a showcase of nature conservation. At a time when the entire country was under military occupation, saving Steller's albatross became an opportunity to show the world that Japan was being reborn: an environmentally responsible society that could be trusted to rejoin the international community of sovereign nations.

Another basic similarity is that performances of colonisation and conservation both relied on a range of strategies for imagining, talking about and manipulating the natural world. These strategies conflicted with each other in obvious ways: put bluntly, were islands to be claimed through killing birds or by saving them? In this sense many disputes over bird islands were political-ecological disputes, as they were contests over how environments were to be managed, by whom and for whose benefit. On the other hand there are also continuities between the two forms of claiming. Both colonisers and conservationists posed as enlightened stewards of the environment, though they had sharply different ideas of what that entailed. For one group stewardship entailed rendering islands habitable, and for another it entailed protecting them from human encroachment. Both turned to their own states to support their goals, and both tapped into wider national anxieties about territorial sovereignty to drum up support for their cause. And both displayed a canny awareness of prevailing dynamics in international affairs, from the high noon of the British Empire to the global territorial scramble of the late nineteenth century, to the US–Japan rivalry in the Pacific, through to the geopolitical posturing that has, to an alarming degree, continued to shape East Asian international relations since World War II. The chief purpose of the book is to trace the connections, dialectical inversions and latent continuities between these two ways of claiming.

The structure of the book is deliberately unorthodox in both periodisation and spatial organisation. Histories of Japan's 'formal' empire conventionally begin with the annexation of Taiwan in 1895 and end with Japan's defeat in World War II. Similarly, histories of US–Japan relations typically start with Commodore Perry's flotilla of Black Ships steaming into Tokyo Bay in 1853. But this study begins in 1830, a thousand kilometres south of that point, and continues the story into the post-war era, and indeed into the early years of the twenty-first century. And although the chapters in this book are arranged in broadly chronological order, they can also be divided thematically: chapters 1, 2 and 4 explore the relationship between colonisation and sovereignty, while chapters 3, 6 and 7 explore the relationship between conservation and sovereignty. Reading the chapters in this order may help bring the dialectical relationship between these two concepts into clearer focus.

In the same vein, readers with particular interests may wish to focus their attention on particular chapters. Chapters 1 and 2 contain the most detailed theorisation of the nineteenth-century Pacific as an 'ocean borderland'. Chapter 3 offsets the organisational focus around Japan by exploring the American reaction against Japanese bird-hunting in and around the Hawaiian archipelago. Those interested in the strategic importance of bird islands should consult chapters 4 and 5. Chapter 6 uses bird islands as a departure point from which to explore the international politics of nature conservation in post-war Japan, while those primarily interested in the contemporary importance of bird islands (both in East Asia and elsewhere) may gain the most from Chapter 7 and the Epilogue.

A Bird's-Eye View of History?

Finally, a word or two about what is *not* in this book. While the archive is replete with accounts furnished by scientists, entrepreneurs and bureaucrats, most of those who lived and worked on bird islands left no written record of their experiences there. Some argue that, as the absence of these voices from the archive is itself a product of power relations, historians have an ethical obligation to reconstruct subaltern perspectives by reading archives 'against the grain'. Others warn that to do so risks appropriating the voices of the powerless to serve the writer's own historiographical or political agenda. While I am sympathetic to both approaches, here I have generally erred on the side of caution, even if the end result admittedly furnishes only a partial view of history. So while I have described the environments and working conditions faced by bird-hunters and guano miners, I have generally stopped short of speculating about their lived experience on bird islands. Readers are, of course, more than welcome to fill in the gaps themselves should they choose.

By the same token, any study of the Pacific must engage with the presence (or absence) of Pacific Islanders in the historical record. Many classic studies of Japanese imperialism have been justly criticised for marginalising Indigenous Taiwanese and Micronesian voices, in large part because they have failed to critique the silences that exist in colonial archives. I have certainly been wary of repeating this mistake. So I want to make clear that although this book mainly focuses on those islands that do *not* have a long history of Indigenous habitation, the vast majority of Pacific islands very much *do*. Indeed, readers who wish to learn more about the history of the Indigenous Pacific are spoiled for choice.[34] Where Pacific Islanders do *not* appear in this story (they play only a marginal role in the chapter on post-annexation Hawai'i, for instance), I can only assure the reader that is not for want of searching. And although there are grounds for arguing that Japanese migrants to Hawai'i were complicit in settler-colonial expropriation of Indigenous lands, the evidence suggests that Native Hawaiians were no longer frequenting outlying islands such as Midway (Hawaiian: Kuaihelani) or Laysan (Kamole) by the time that Japanese bird-hunters began visiting them.[35]

Secondly, I have taken pains to highlight those instances in which Pacific Islanders did play a role in the story. The majority of early settlers on the Bonin Islands, for instance, were Polynesian, and Anglo attitudes to interracial marriage do much to explain British and US diplomacy in the nineteenth-century Pacific. Elsewhere I have drawn on oral accounts of the Pacific War to provide a Micronesian perspective on the Battle of Saipan. On both the Bonins and Saipan islander agency was at best Schweikian (as James C. Scott puts it),

[34] Nicholas Thomas, *Islanders: The Pacific in the Age of Empire* (New Haven: Yale University Press, 2010); Matt K. Matsuda, *Pacific Worlds: A History of Seas, Peoples, and Cultures* (Cambridge: Cambridge University Press, 2012); David Chang, *The World and All the Things upon It: Native Hawaiian Geographies of Exploration* (Minneapolis: University of Minnesota Press, 2016).

[35] Dean Itsuji Saranillio, 'Why Asian Settler Colonialism Matters: A Thought Piece on Critiques, Debates, and Indigenous Difference', *Settler Colonial Studies* 3 no. 3–4 (2013): 280–94. Kekuewa Kikiloi's research shows that until the early nineteenth century the Northwestern Hawaiian Islands 'were considered 'äina akua, or ancestral islands where the souls of the deceased would travel to and lived in afterlife.' But such beliefs appear to have faded with the collapse of the kapu system and the adoption of Christianity. They are now being reconstituted, in large part thanks to Kikiloi's own efforts. See Kikiloi, 'Rebirth of an Archipelago', 89.

manifest in covert tactics of resistance such as flight, feigned incomprehension and dissemblage.[36] In many ways it is difficult to avoid the impression that the islanders were often victims of powerful forces beyond their control. But although it would be difficult to read this book as a celebration of islander agency, neither is it intended as a denial of islander humanity.[37]

Talk of humanity leads on to another point. Recent years have seen numerous spirited attempts to incorporate non-human perspectives into the writing of environmental history.[38] I am sympathetic to this effort to 'expand the circle' of compassion, and certainly a great many animals died in the course of the story I am about to tell, some in great pain.[39] But in the end I chose to adopt a stodgily anthropocentric perspective. This was partly for practical reasons. Just as with human subalterns, the voices of animals yawn like an abyss in the archival record. But there are also inherent philosophical problems related to the limitations of language in conveying lived experience. To paraphrase Wittgenstein, if an albatross could talk we could not understand it.[40] And recent research into the cognitive faculties of crows, cephalopods and fungi underscores the profoundly alien character of non-mammalian consciousness in particular.[41] Though scientists can say with confidence that consciousness is possible in organisms that lack a cerebral cortex, even speculating on the nature of that consciousness is for me a bridge too far.[42]

36 James C. Scott, *Weapons of the Weak: Everyday Forms of Peasant Resistance* (New Haven: Yale University Press, 1987), xvi; Michel Foucault, 'The Eye of Power', in *Power/Knowledge: Selected Interviews and Other Writings, 1972–1977* (New York: Vintage, 1980), 69.

37 Walter Johnson, 'On Agency', *Journal of Social History* 37 no. 1 (2003): 115.

38 Walker, *The Lost Wolves of Japan*; Gregory T. Rosenthal, 'Life and Labor in a Seabird Colony: Hawaiian Guano Workers, 1857–70', *Environmental History* 17(2012): 744–82; Bathsheba Demuth, *Floating Coast: An Environmental History of the Bering Strait* (New York: W. W. Norton & Co., 2019).

39 Peter Singer, *The Expanding Circle: Ethics, Evolution, and Moral Progress* (Princeton: Princeton University Press, 2011).

40 Kari Weil, *Thinking Animals: Why Animal Studies Now* (New York: Columbia University Press, 2012), 7.

41 Peter Godfrey-Smith, *Other Minds: The Octopus, the Sea, and the Deep Origins of Consciousness* (William Collins, 2016); Merlin Sheldrake, *Entangled Life: How Fungi Make Our Worlds, Change Our Minds and Shape Our Futures* (Random House, 2020).

42 Andreas Nieder, Lysann Wagener and Paul Rinnert, 'A Neural Correlate of Sensory Consciousness in a Corvid Bird', *Science* 369 no. 6511 (25 September 2020): 1626–9.

Finally, as theorists of post-humanism have pointed out, any attempt to include non-human perspectives in a given narrative almost invariably runs into the thorny question of exactly *which* perspectives to privilege.[43] Innumerable lifeforms coexist on and around islands in intricate, often unstable webs of life: not only birds but also rats, goats, pigs, squid, coral, sand-flies, enteric bacteria and a whole host of others.[44] To emphasise the perspective of a particular species might risk putting my finger on the scales, insofar as it could be interpreted as asserting the moral value of that species' lives over the lives of others.

I am particularly wary of echoing the biases of my sources: those naturalists who were solely concerned with the fates of 'native' fauna, to the exclusion of 'invasive' species such as rats, pigs and goats. This danger is especially acute when it comes to the albatross, a seductively charismatic creature that in the Anglo imagination has deep, semi-mystical connections to the notion of pristine wilderness. The trouble with wilderness, as Cronon has pointed out, is that it obscures the inherently political nature of nature: that different actors often have very different ideas about what the natural world should look like.[45] So any attempt to ventriloquise albatrosses might easily be misconstrued as a claim to speak on behalf of nature itself.

Bruno Latour takes the argument a step further, suggesting that political ecologists talk not of one single nature but of multiple possible natures, none of which has any inherent moral value over the others.[46] In this spirit, I have resisted the temptation to provide a qualitative assessment of the environmental change wrought by bird-hunters, guano miners and conservationists. It is undeniable that the age of extraction changed the ecology of the Pacific drastically, with human activity depleting stocks of biomass accrued over millennia within a few short years – virtually the blink of an eye within geological time.[47] More recent trends have been documented extensively elsewhere: the decimation of fishery stocks and the saturation of the oceans

43 Weil, *Thinking Animals*, 25.
44 Victoria Braithwaite, *Do Fish Feel Pain?* (Oxford: Oxford University Press, 2010).
45 Cronon, 'The Trouble with Wilderness'.
46 Bruno Latour, *The Politics of Nature: How to Bring the Sciences into Democracy* (Cambridge, MA: Harvard University Press, 2004), 29.
47 John. R. McNeill, 'Of Rats and Men: A Synoptic Environmental History of the Island Pacific', *Journal of World History* 5 no. 2 (1994): 312.

with plastic waste, manifest most vividly in 'the Great Pacific Garbage Patch'.[48]

But it is not the project of this book to evaluate the Pacific's transition from a prelapsarian state of nature to an anthropocene one. Rather, it is to show the conditions under which particular ideas about how the natural world ought to be governed came to take hold. Some ecologists today stress the value of albatrosses as 'apex predators' whose faeces stimulate phytoplankton growth, thereby helping to keep entire terraqueous ecosystems in balance.[49] This may very well be true, but I do not believe it to be relevant to the study at hand. Indeed, to place undue stress on this point risks muddying the waters, for it is difficult to analyse the social and material conditions under which knowledge of the natural world is produced while at the same time also endorsing one form of knowledge over others. On the contrary, writing an effective political ecology of Japan's ocean borderlands requires a degree of analytical distance, so that we may unpack the political uses of science without succumbing to the same temptation ourselves. Nature, in other words, is an albatross around our necks.

[48] Michelle Allsopp et al., *State of the World's Oceans* (Dordrecht: Springer, 2010); L. Lebreton et al., 'Evidence That the Great Pacific Garbage Patch Is Rapidly Accumulating Plastic', *Nature Scientific Reports* 8 no. 1 (22 March 2018).

[49] Matthew S. Savoca and Gabrielle A. Nevitt, 'Evidence that Dimethyl Sulfide Facilitates a Tritrophic Mutualism Between Marine Primary Producers and Top Predators', *Proceedings of the National Academy of Sciences of the United States of America*, 111 no. 11 (2014): 4157–61.

1 Bonins of Contention

Extraterritorial Empire and Borderland Citizenship in the Nineteenth-Century Pacific

We begin with a tussle over a gun, six hundred miles out to sea. It was 1863, and a motley crew of whalers had hauled anchor off the uninhabited island of Anijima to flense a sperm whale. Processing a dead whale is gruelling work. Its body must be lashed alongside ship and its blubber sliced into chunks, then boiled in an iron pot to extract the oil. The crew, already tired from hours of pursuing the whale in oar-powered longboats, must work quickly before the floating carcass is devoured by sharks.[1] So tempers aboard the *Ichiban Maru* would already have been frayed when the captain accused one of his crew members of shirking on the job, and confiscated the man's belongings as punishment. An everyday quarrel over compensation suddenly escalated, however, when the captain discovered that a pistol had been smuggled aboard ship. He promptly accused the man and his friend of piracy, ordered his crew to arrest the suspects, and hauled anchor in search of some higher authority who might resolve the dispute.[2]

Who that higher authority might be was not immediately obvious, for the crew of the *Ichiban Maru* consisted of a jumble of differing nationalities thrown together into the same boat by the same sweeping global processes. The exact location of the dispute only complicated matters still further. Anijima was one of the Bonin Isles (J: Ogasawara Shotō), a handful of tiny islands scattered deep in the Pacific Ocean. The Bonins had, until 1830, only the most fleeting history of human habitation. Indeed, their very name derives from a Japanese word meaning 'unpeopled isle[s]'.[3] Yet by 1863 they had emerged as a small but vital provisioning hub within a vast trans-oceanic extractive economy. And the Bonin population, though few in number, was global in composition – hailing from islands across the Pacific

1 Demuth, *Floating Coast*, 42–3; Richard Ellis, *Men and Whales* (New York: Lyons Press, 1999), 34–8.

2 Ishihara, *Kindai Nihon to Ogasawara Shotō*, 204.

3 *bunin-jima*. This archaic pronunciation of 無人島 differs from the modern word *mujintō*, a generic term for uninhabited islands.

but also from Japan, America and Europe. This mix of nationalities created what Ziad Fahmy has called a 'jurisdictional borderland' where, in the absence of established territorial sovereignty, a number of different governments strove to extend economic and legal protection to their citizens.[4] And the *Ichiban Maru* may have been captained by a Japanese national under a Japanese flag, but the two accused crew members both claimed citizenship of the United States, an aspiring imperial power known for asserting its Pacific interests down the barrel of a gunboat.

The Bonin Islands thus make an ideal vantage point from which to survey the oceanic borderland of the North Pacific in the nineteenth century. How did the people who lived and laboured in this borderland space negotiate with their environment and with each other? How did they resolve the inevitable disputes that arose during a time of intense economic disruption and political instability? What forces drove people to colonise remote specks of storm-battered rock so far out to sea? And what factors contributed to the islands' eventual transformation from borderland space to the outermost territory of the Japanese state?

To answer these questions I first trace the rise and fall of the global whaling industry, detailing the global conditions that made it possible and the effects that it wrought on environments, societies and people throughout the North Pacific. I then switch focus to examine the local histories that brought four different groups of settlers – Hawaiians, beachcombers, mainland Japanese and Hachijōjimans – to the Bonin Islands by the mid-nineteenth century. This set the stage for the *Ichiban Maru* incident, a petty dispute that soon spiralled into an international confrontation between Japanese, US and British diplomats. Finally, I explore the historical forces that transformed the Bonins from borderland space to bordered space: Japanese settler colonialism, yes, but also a calculated decision on the part of Anglo diplomats to slough off responsibility for a new, racially mixed generation of Bonin Islanders.

Global Whaling in the Pacific: Mobility, Empire and Extraterritoriality

Whaling was the earliest and most notorious of the many extractive industries that dominated the economy of the nineteenth-century Pacific (Figure 1.1).[5] It also foreshadowed later activities such as bird-hunting and

4 Fahmy, 'Jurisdictional Borderlands'. 5 McNeill, 'Of Rats and Men', 312.

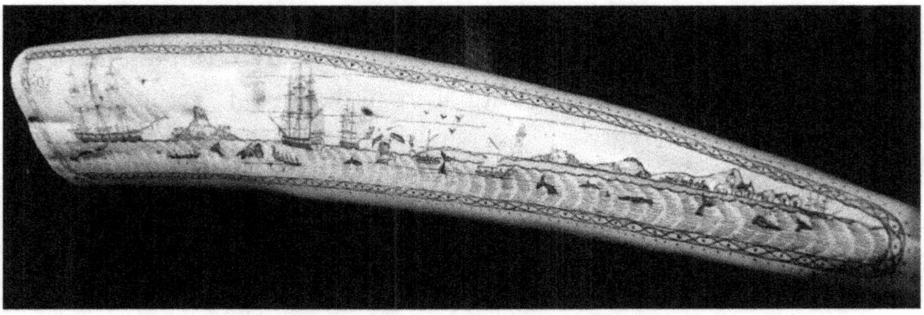

Figure 1.1 Scrimshaw etching of Arctic whaling expedition on walrus tusk (n.d.)
Source: National Museum of Ethnology collection, Osaka

guano mining in important ways, most notably in its extreme mobility, its vulnerability to boom and bust cycles, and its lack of regard for any kind of sustainable resource management. Moreover it was, at the time, something fundamentally new. Japanese and Indigenous people throughout the Pacific have hunted whales for centuries but from around 1780, after whale sightings in the Atlantic became scarce, British, French and American whalers began venturing round Cape Horn into the Pacific.[6] This represented both a step-change in scale and a profound transformation in the way the industry was organised. The actual techniques used to hunt did not change much at first, but sending an expedition half-way around the world on a voyage involved raising substantial venture capital funds, outfitting a much larger vessel, and recruiting captains and crew who knew how to sail the open ocean.[7] The end purpose of the expeditions was also quite different: while earlier hunters had generally made use of all parts of the whale, global whalers jettisoned most of the carcass and focused exclusively on extracting oil and baleen.[8] The former was turned into lamp oil through a complex industrial process of 'pressing, fluxing

6 Jakobina Arch, *Bringing Whales Ashore: Oceans and the Environment of Early Modern Japan* (Seattle: University of Washington Press, 2018); Randall R. Reeves, 'The Origins and Character of "Aboriginal Subsistence" Whaling: A Global Review', *Mammal Review* 32 no. 2 (June 2002).

7 Eric Dolin, *Leviathan: The History of Whaling in America* (New York: W. W. Norton, 2007), 181; Lance E. David, Robert E. Gallman and Karin Gleiter, *In Pursuit of Leviathan: Technology, Institutions, Productivity, and Profits in American Whaling, 1816–1906* (Chicago: University of Chicago Press, 1997), 294–5; Tom Nicholas, *VC: An American History* (Cambridge, MA: Harvard University Press, 2019), 12–13, 23.

8 McNeill, 'Of Rats and Men', 320–1; Ryan Tucker Jones, 'Running into Whales: The History of the North Pacific from Below the Waves', *The American Historical Review* 118 no. 2(April 2013): 359; Arch, *Bringing Whales Ashore*, Ch.3.

and crystallising'; the latter was fashioned into whalebone corsets and other accoutrements.[9] The exclusive focus on these two commodities does much to explain the boom and bust nature of the global whaling industry. The increasing availability of cheaper, cleaner-burning kerosene by the mid-nineteenth century sent demand for whale oil into a permanent decline, with catastrophic effects on the profitability of the industry as a whole.[10]

Global whaling also relied on a fundamentally different labour regime. Local whaling had been conducted at the level of individual fishing villages on a seasonal or simply opportunistic basis.[11] But the crews of global whaling expeditions worked under terms that were close to indentured labour. Many were pressed or blackbirded against their will into multi-year stints onboard, with no guarantee that they would be returned home at the end of it. Even those whalemen who signed up voluntarily sometimes chose to jump ship along the way rather than endure the harsh working conditions. And because whaling crews earned their compensation as a share of the profit earned by the expedition as a whole, many ended up penniless or even in debt at the end of their voyage. The result was to create a kind of trans-Pacific maritime proletariat who drifted unmoored from port to port.

The sheer scale of global whaling also demanded much more elaborate logistical networks. Ships needed to restock on food, water and fuel, a need that became particularly pressing during long oceanic crossings where ports of call were scarce. And because expeditions lasted between three and four years on average, it became routine for captains to send their cargo home mid-voyage. This required secure transshipment networks, so that the harvested oil and baleen could arrive safely back at port without being lost or pilfered en route.[12] It required, in other words, a global regime that enforced private property rights and guaranteed the unimpeded flow of goods and people – what Robinson and Gallagher famously called an 'imperialism of free trade'.[13]

9 Dolin, *Leviathan*, 110, 356. 10 Demuth, *Floating Coast*, 48–9.

11 Arch, *Bringing Whales Ashore*, 50; Noell Wilson, 'Whaling at the Margins: Drift Whales, Ainu Laborers, and the Japanese State on the Nineteenth-Century Okhotsk Coast', in *New Histories of Pacific Whaling*, ed. Ryan Tucker Jones and Angela Wanhalla (Munich: Rachel Carson Center for Environmental History, 2019), 57–64.

12 Nicholas, *VC*, 18; David, Gallman and Gleiter, *In Pursuit of Leviathan*, 46.

13 John Gallagher and Ronald Robinson, 'The Imperialism of Free Trade', *The Economic History Review New Series* 6 no. 1 (1 January 1953): 1–15.

Techniques of gunboat diplomacy that were later used to force open Chinese, Japanese and Korean ports were earlier deployed to protect the whalers that called at islands throughout the oceanic Pacific.

These military interventions did not aim at territorial aggrandisement; rather, they constituted a kind of extraterritorial imperialism that asserted the right to selectively protect certain types of people and property. As Lauren Benton puts it, they did not so much 'corral law into conventionally defined jurisdictions' as create 'repeating sets of irregularly shaped corridors with ambiguous and shifting relations to imperial sovereignty'.[14] And they triggered profound political, social and economic ructions throughout the Pacific.

In the following two sections we shall explore the effects of this extraterritorial imperialism on two different countries, Hawai'i and Japan. We shall see how the whaling industry and the gunboat diplomacy that supported it destabilised existing political structures and, at the same time, helped create a new kind of oceanic contact zone in the North Pacific. Human settlement on the Bonin Islands was a part of this new borderland space. And it was governed to a particularly stark degree by the same extraterritorial logic that shaped the rest of the Pacific, East Asia and the world.

From Hawai'i to the Bonins

Hawai'i might seem a quixotic place to begin a history of Japan's borderlands, but there are good reasons to do so. For one thing, the majority of the first Bonin settlers hailed from Hawai'i. Secondly, in the mid-nineteenth century Hawai'i and Japan were in somewhat similar positions. In later years the fates of the two countries would diverge dramatically – one was colonised by a more powerful neighbour, and the other went on to become an imperial power in its own right – but at the time our story begins both were relatively weak states within a rapidly globalising Pacific world. (In addition, the events of Chapter 3 substantially play out in post-annexation Hawai'i, so a brief pre-history here may help provide relevant background context.)

In the decades following the encounter with Captain Cook, the Kingdom of Hawai'i had become thoroughly interwoven in global webs

[14] Lauren Benton, *A Search for Sovereignty: Law and Geography in European Empires, 1400–1900* (Cambridge: Cambridge University Press, 2009), xii.

of commerce, resource extraction and empire.[15] King Kamehameha I had succeeded in unifying the archipelago, in large part using weapons and mana (supernatural power) supplied by the British, and the new monarchy had prospered.[16] It had secured pledges of loyalty from prominent political figures, including some foreigners.[17] It had exploited Hawai'i's position on North Pacific whaling and trading routes to turn the kingdom into a provisioning hub for passing ships. And it had coordinated the felling of the islands' sandalwood groves for export to China as incense.[18] It had used this revenue to purchase top hats, frocks and all the accoutrements of Western civilisation, and to despatch a mission to London to secure recognition from the paramount imperial power of the time.[19]

This diplomatic recognition almost certainly helped to protect Hawai'i from outright annexation, at least up until 1898. But it did not translate into full territorial sovereignty. On the contrary Hawai'i, like many other Pacific islands, remained vulnerable to attack by any warship that happened to pass through. As early as 1806 Britain passed an act 'extending jurisdiction over crimes committed under its flag on the high seas to crimes committed by its subjects on Pacific Islands ... Naval warships were to arrest British miscreants and inflict war-like punishment on island communities for destroying British property and massacring British crews.'[20] In 1826 King Kamehameha III issued an edict to the effect that foreigners should be subject to Hawaiian law, but it was widely flouted, and by 1840 combined pressure from Britain, France and the United States had succeeded in codifying rather haphazard techniques for enforcing extraterritorial privilege into a system of trial by consular court.[21]

15 Chang, *World and All the Things*, 33–35.

16 Juri Mykkänen, *Inventing Politics: A New Political Anthropology of the Hawaiian Kingdom* (Honolulu: University of Hawai'i Press, 2003), 38.

17 Stuart Banner, 'Preparing to be Colonized: Land Tenure and Legal Strategy in Nineteenth-Century Hawaii', *Law & Society Review* 39 no. 2 (2005): 283.

18 Ralph Kuykendall, *The Hawaiian Kingdom Volume I: Foundation and Transformation, 1778–1854* (Honolulu: University of Hawai'i Press), 125. Banner echoes ninteenth-century European observers in describing early Hawai'i as a quasi-feudal society. Banner, 'Preparing to be Colonized', 281.

19 Coll Thrush, *Indigenous London: Native Travelers at the Heart of Empire* (New Haven: Yale University Press, 2016), 151.

20 Lisa Ford, 'Law' in *Pacific Histories: Ocean, Land, People*, eds. David Armitage and Alison Bashford (Basingstoke: Palgrave Macmillan), 223.

21 Juri Mykkänen, *Inventing Politics: A New Political Anthropology of the Hawaiian Kingdom* (Honolulu: University of Hawai'i Press, 2003), 38; Kuykendall, *Hawaiian Kingdom* 1, 129, 166–7.

During the same period, huge numbers of Hawaiians left their home island and ventured abroad. By 1846 an astonishing one in five men between the ages of fifteen and thirty were living outside Hawai'i, crewing whalers, mining guano on desert islands or panning for gold in the California desert.[22] But these expatriate Hawaiians did not enjoy the same protection as the average British or American sailor on shore leave in Honolulu. Though Hawai'i actually boasted a remarkably widespread network of consular officials, no Hawaiian (or Tahitian or Fijian) warship could intervene to defend the rights of an islander accused of manslaughter in a New Bedford bar brawl – for no such warships existed.[23]

Such was the vulnerability of the Hawaiians who, from 1830, began to settle on the Bonin Islands. Frustratingly little survives in the historical record to recover the perspective of these early colonists. But we know they spent the first months on the Bonins clearing jungle from the islands' few pockets of flat land, subsisting on the pith of cabbage trees and the meat of the turtles that swam from across the Pacific to lay their eggs on the islands' beaches. We also know that they introduced a portmanteau of flora and fauna into the islands, much as their ancestors had done elsewhere in Polynesia for thousands of years.[24] (Indeed, the Hawaiian and Bonin archipelagoes lie at similar latitudes, and have similar climate and geology.)[25] We know that they grew sweet potato, taro, yams and other root vegetables, and ate them by mashing them up into poi, a Polynesian culinary staple.[26] And that they also grew Indian corn, pumpkins, watermelon, sugar cane, lemon trees and tobacco. And we know that they fished using Polynesian-style outrigger canoes, that they kept chickens and hogs, that they permanently transformed the ecology of the islands by releasing goats to graze the

22 Gregory T. Rosenthal, *Beyond Hawai'i: Native Labor in the Pacific World* (Oakland: University of California Press, 2018), 77.

23 By 1887 the Kingdom of Hawai'i 'maintained 103 legations and consulates worldwide'. That same year Hawai'i did in fact purchase a former British trading schooner and outfit it as a naval ship, but 'its impression on the Western powers with their multiple ironclad warships in the region could only be symbolic'. See Gonschor, *A Power in the World*, 37, 98.

24 McNeill, 'Of Rats and Men', 304–5.

25 Nanyan Guo, 'Environmental Culture and World Heritage in Pacific Japan: Saving the Ogasawara Islands', *The Asia-Pacific Journal: Japan Focus* 7 issue 17 no. 3 (12 April 2009).

26 Scott Kramer and Hanae Kurihara Kramer, 'The Other Isles of Aloha', *The Journal of Hawaiian History* 47 (2013): 7, 12.

hillsides and that they even trained dogs to catch white-tipped sharks in the shallow water off the islands' long sandy beaches.[27] Hawaiian language and culture permeated life in the new colony: Japanese castaways who washed up on the Bonins in 1839 described the islanders as dressing in the Hawaiian style and speaking a mixed Hawaiian-English pidgin.[28] All of which is to say that these early settlers did a huge amount to imprint Hawaiian society and ecology onto the islands during the early years of the colony (Figure 1.2).

Hawaiians were not the only Pacific Islanders who settled on the Bonins. Over time people from Saipan, Guam and as far afield as the Marquesas Islands came to live on the islands. This was a function of the extreme mobility that characterised life on the islands in these early years, for many of the new arrivals alighted from passing whalers, frigates and merchantmen. Indeed, the settlement on the Bonin colony was not so much a self-sufficient agricultural community as a maritime provisioning depot.[29] Merchant vessels plying the trade route between China and the Pacific Northwest stopped off to stock up on food, water and timber, and ships from as far afield as Dunkerque, New Bedford and Auckland hunted the sperm, right and humpback whales that migrated along the 30th parallel, feeding on the plankton that welled up from the ocean into the waters surrounding the islands.[30]

This constant trans-oceanic traffic provided opportunities to barter provisions such as firewood and turtle meat, and to charge pilot fees to navigate the islands' perilous reefs. It also enabled the islanders, already migrants at least one time over, to continue living startlingly mobile lives. Some signed up for stints aboard whalers or sealing expeditions to the Kurile Islands, for example.[31] Others wearied of life on the islands and left altogether, never to return. They were replaced by new arrivals from elsewhere, so that the population of the islands was constantly in flux to the point at which the distinction between visitor, sojourner and permanent colonist became a blurry one.

27 David Chapman, *The Bonin Islanders, 1830 to the Present: Narrating Japanese Nationality* (Lanham: Lexington Books, 2016), 30–1. On the use of outrigger canoes in the Bonins, see NA FCO 46/195 Robertson to Parkes: 141.

28 Kramer and Hanae, 'The Other Isles of Aloha', 17.

29 Chapman, *Bonin Islanders,* 29–30.

30 Tim D. Smith, Randall R. Reeves, Elizabeth A. Josephson and Judith N. Lund, 'Spatial and Seasonal Distributions of American Whaling and Whales in the Age of Sail', *PloS One* 7 no. 4 (April 2012): 3.

31 Ishihara, *Kindai Nihon to Ogasawara-tō*, 287.

(a)

(b)

Figure 1.2a Outrigger canoe used by Bonin Islanders (c.1862)
Figure 1.2b Bonin Islander women dancing on Hahajima (c.1862)
Source: Obana Sakusuke, *Ogasawara-jima zu-e* (Tokyo: Ogasawara-mura Kyōiku Iinkai, 2009). Ogasawara-mura Education Committee collection

Pacific Islanders and Beachcombers

Another type of person lived alongside the Hawaiians in this early Bonin settlement: beachcombers.[32] Beachcombers were ubiquitous in the nineteenth-century Pacific. Almost always male, they hailed from Europe or the Neo-European settler colonies but spent large stretches of their lives drifting from port to port trying their hand at fur-sealing, whaling or stevedoring, or trading in sundry goods such as rum, textiles or guns.[33] In their extreme mobility beachcombers resembled Hawaiians, and indeed they often worked alongside them, doing similar work in similar conditions. Some beachcombers even ingratiated themselves within Polynesian chiefly households – many pledged oaths of loyalty to the Hawaiian monarchy, for example.[34] But at the same time, beachcombers took care to assert their citizenship of those states that had any kind of naval presence in the Pacific – Britain, of course, but also France, Russia and the US – mindful of the extraterritorial protection this might afford. This set them quite apart from the Hawaiians, for they were the beneficiaries of the very same extraterritorial imperialism that constrained Hawaiian sovereignty during this period. They were politically (if not always ethnically) white.

From the outset on the Bonin Islands, beachcombers were substantially outnumbered by Hawaiians. Yet they exploited their privileged status in order to style themselves as the leaders of the new settlement. Two British citizens, Millichamp and Mazzaro, had secured cautious support for the colonisation venture from the British consul at Honolulu prior to departure. They spoke of their Hawaiian co-settlers as their 'servants' or even 'slaves'; claims corroborated by a visiting British sea captain who, in 1837, reported that the Hawaiians had been bound to work for the Britons for a period of years before earning the right to live freely on the islands.[35] A second

[32] For a discussion of 'beachcombers' versus 'kanaka' see Ishihara Shun, *'Guntō' no rekishi shakaigaku: Ogasawara Shotō, Iwōtō, Nihon, America, soshite Taiheiyō Sekai* (Tokyo: Kōbundō, 2013) 40–2. For the European 'seizure' of the word kanaka see Chang, *World and All the Things*, 35.

[33] On Neo-European settler colonies see Alfred W. Crosby, *Ecological Imperialism: The Biological Expansion of Europe, 900–1900* (Cambridge: Cambridge University Press, 2004).

[34] Thomas, *Islanders*, 49.

[35] Chapman, *Bonin Islanders*, 29–31. Indeed, by the 1860s at least some Hawaiians/Pacific Islanders had managed to establish themselves as

faction was led by Nathaniel Savory, a Massachusetts-born American who, much like Millichamp and Mazzaro, looked to his own government to underpin his authority. In 1853, a flotilla of American warships under the command of one Commodore Matthew Perry called at Peel Island (J: Chichijima), the main site of the largest Bonin settlement. Perry had been charged with opening up a steamer route between California and China, and Savory was able to convince him that the Bonins would make an ideal coaling station.[36] The commodore conducted a cadastral survey as a prelude to annexation by the United States, and before he left pressured the islanders to ratify a provisional constitution that appointed Savory as headman.[37]

Negotiations such as these produced on the Bonins a society that in many resembled the broader Pacific world. It consisted of two classes, beachcombers and Pacific Islanders, politically distinct insofar as the former could use their citizenship to seek the protection of vastly more powerful governments than the latter.[38] That said, this protection did not always materialise. Even at the height of the Pacific whaling industry the Bonin Islands remained isolated from the outside world for large stretches of time. Fragmentary external accounts suggest that the islanders often resolved their disputes by relying upon what Ishihara Shun calls 'local law', and what other, less sympathetic observers called 'club law' or 'lynch law'.[39] The islanders were also often at the mercy of the very sailors upon whom they relied for their living. On at least two occasions, in 1835 and 1842, the British consul at Honolulu wrote to London on behalf of the islanders, 'praying for protection against vessels employed in the whale fishery'.[40] In one notorious incident in 1849, whalers ransacked the islands for livestock, food supplies and other stores, as well as relieving Savory of two thousand dollars and his

independent cultivators. See Katsu Kaishū, *Kaigun rekishi* 5 no. 11 (Tokyo: Kaigunshō), 26–7, 31–8.

36 Webster to Aulick (10 June 1851) cited in 'Japan–the Expedition', *The American Whig Review* 15 (June 1852), 514.

37 NA FCO 46/195 Robertson to Parkes (13 December 1875): 124.

38 Francis L. Hawks, *Narrative of the Expedition of an American Squadron to the China Seas and Japan Performed in the Years 1852,1853 and 1854 Under the Command of Commodore M.c. Perry, United States Navy, By Order of the Government of the United States* (Washington, DC: United States Congress, 1856), 203.

39 Ishihara, *Kindai Nihon to Ogasawara-tō*, 240; NA FCO 46/195 Robertson to Parkes (27 December 1875), 123.

40 Chapman, *The Bonin Islanders*, 29, 35.

Hawaiian wife (who may have been a willing conspirator in the whole affair).[41]

It is also the case that, privately at least, officials in London and Washington, DC, were often indifferent to the fate of their citizens on the Bonin Islands. Although a British naval captain had in fact made landfall on the islands in 1827 and claimed them for the British Empire, Colonial Office bureaucrats declined to ratify the annexation.[42] Neither did the US State Department ever officially annex the islands – Perry's enthusiasm notwithstanding. As for combating piracy, British officials responded to their own consul's entreaties with the bland observation that 'the island of the Bonin is beyond the limits to which British cruisers ordinarily go'.[43] The US Secretary of State was even more scathing, writing of his compatriots that 'by resorting to such remote spots on the globe's surface, under such circumstances, they may fairly be held to have deliberately abandoned the United States without purpose of returning, and therefore to have relinquished the rights as well as duties of citizens'.[44] Comments such as these reveal that the protection afforded by beachcomber citizenship did not always materialise. It often existed as a hope – or a threat.

Japan in the Extraterritorial Pacific

After Perry left the Bonin Islands, he let the water of the Kuroshio current sweep his ships north towards another promising site for a coaling station. Japan was in many ways very different to Hawai'i, but in other ways similar. Japan was, of course, larger and more populous than Hawai'i by orders of magnitude. And by the time Hawai'i was unified in 1795, Japan had enjoyed nearly two centuries of uninterrupted peace under the rule of the Tokugawa Shogunate. Whereas Hawai'i was largely cut off from the outside world between the initial Polynesian colonisation and the arrival of Captain Cook, Tokugawa Japan maintained diplomatic, economic and cultural ties with its neighbours. Indeed, the Tokugawa were accustomed to conducting their diplomacy from a position of strength, making a great show of welcoming tribute missions from Korea, the Ryūkyū

41 *Ibid*, 36.
42 Frederick William Beechey, *Narrative of a Voyage to the Pacific and Beering's Strait* (Amsterdam: U.A. Israel, 1968), Ch.6.
43 Chapman, 29–30. 44 *Ibid*, 79.

Kingdom, the Ainu, the Khmer, and even the Dutch East India Company.[45]

Yet for all these differences, the Tokugawa Shogunate eventually found itself enmeshed in the same extraterritorial imperialism that had ensnared the Kingdom of Hawai'i, for the same decentralised political structure that had kept the peace domestically also rendered Japan vulnerable to the vicissitudes of gunboat diplomacy. Its political system was highly decentralised, with large swathes of the country ruled by powerful domainal lords who enjoyed a high degree of autonomy. And the Shogunate possessed no real navy to speak of, beyond a few coastal patrol boats tasked with intercepting smugglers.[46] In 1808 a British frigate sailed into the port of Nagasaki and extorted food and supplies from the harbourmaster at cannon-point – exposing the country's lack of military preparedness in the face of the latest naval technology.[47] By the 1820s whalers flying British and American flags were being regularly spotted not only along the Pacific coastline but deep into the Sea of Japan. Some called at coastal fishing villages to secure provisions, and when one of these encounters descended into violence the Shogunate implemented a 'shell-and-repel' policy in the hope of deterring future interactions.[48] The policy was abandoned in 1842, however, after word reached Japan of Qing China's defeat against Britain in the First Opium War.[49] And when Commodore Perry's flotilla of warships steamed into Edo Bay the Shogunate decided that discretion was the better part of valour and accepted his demand to open up ports to American merchants.

Within five years the Shogunate had negotiated and signed treaties not only with the United States but also with Britain, France, Russia and the Netherlands. These treaties granted similar concessions to

45 Ronald Toby, 'Carnival of the Aliens: Korean Embassies in Edo-Period Art and Popular Culture', *Monumenta Nipponica* 41 no. 4 (1986): 415–56; Mark Ravina, *To Stand with The Nations of the World: Japan's Meiji Restoration in World History* (New York: Oxford University Press, 2017), 60; Travis Seifman, 'Performing "Lūchū": Identity Performance and Foreign Relations in Early Modern Japan' (PhD diss., University of California Santa Barbara, 2019).

46 Noell Wilson, *Defensive Positions: The Politics of Maritime Security in Tokugawa Japan* (Cambridge, MA: Harvard University Press), 67.

47 *Ibid*, 80, 115–18.

48 David L. Howell, 'Foreign Encounters and Informal Diplomacy in Early Modern Japan', *The Journal of Japanese Studies* 40 no. 2 (Summer 2014): 309.

49 Ravina, *To Stand with the Nations of the World*, 67.

those granted by Hawai'i a few years earlier, and also by Qing China, the Kingdom of Siam, the Ottoman Empire and many other states that found themselves drawn into the ambit of extraterritorial imperialism during this period.[50] The most galling concessions were the limits on tariffs, which undermined a long-existing ban on bullion exports and would end up playing havoc with local commodity prices. Being forced to allow foreigners to reside in certain Japanese port cities, henceforth known as treaty ports, also grated. Compared to these, the clauses granting consular protection to those foreigners did not initially raise much concern among Tokugawa diplomats. Plural jurisdictions were already central to the Tokugawa legal code: domainal lords exercised broad autonomy within their fiefs, and indeed the entirety of early modern Japanese society was organised into self-governing status groups whose leaders assumed group responsibility for punishing their members according to the Shogun's legal code. Shogunal officials assumed that the consular courts mandated in the new treaties would function as just another vessel in this 'container society'.[51]

Belatedly, however, the officials realised that consular courts were not simply institutions for enforcing judgements handed down by the Shogunate; they had the power to undermine those judgements by ameliorating or even waiving punishment. Interrogation by torture was an integral part of the Tokugawa penal system but consular officials viewed the practice as barbaric; yet trial by jury raised the possibility that the accused might escape sanction altogether. Other quirks of the consular court system also weakened Shogunal authority over foreign nationals. In September 1860, a Briton named Moss flouted a ban on duck-hunting in the vicinity of the Shogun's castle. When confronted by Japanese officials he fired upon them, seriously wounding one of them. Initially the consular courts issued Moss with a stiff fine and ordered him deported to Hong Kong to serve three months of imprisonment. But on 23 January 1861 the Hong Kong courts overruled this decision and ordered Moss released, even awarding damages for wrongful imprisonment into the bargain. To add insult to injury, the foreign merchants'

[50] Turan Kayaoglu, *Legal Imperialism: Sovereignty and Extraterritoriality in Japan, the Ottoman Empire and China* (Cambridge: Cambridge University Press, 2014); Cassel, *Grounds of Judgment*.

[51] Cassel, *Grounds of Judgment*, 30–2, 91–2. For the concept of early modern Japan as a container society see John Whitney Hall, 'Rule by Status in Tokugawa Japan', *Journal of Japanese Studies* 1 no. 1 (Autumn, 1974): 48.

community in Yokohama raised funds to pay Moss' fine for him, so that he effectively got off scot-free for his crime.[52]

When offences were committed *against* foreigners, on the other hand, British diplomats showed little patience for the container society's juridical pluralism. When the British merchant Charles Richardson was slain by retainers of Satsuma Domain in September 1860, the British ambassador demanded reparations from Satsuma; when Satsuma prevaricated, a squadron of British gunboats shelled the domainal capital of Kagoshima until the reparations were paid.[53] Disputes between Japanese and treaty-port-dwelling foreigners thus regularly flared up into crises that roiled both domestic and foreign affairs, and played no small part in hastening the Shogunate's eventual collapse in 1868.

The Tokugawa Colonisation of the Bonins, 1861–1863

All that said, the Shogunate was by no means passive in the years leading up to its demise. On the contrary, it embarked on a number of efforts to combat the foreign threat. To boost the country's coastal defences it began relaxing the strictures on military service, hitherto limited to the warrior caste, by forming commoner militias.[54] It gave domain lords increased autonomy to build up their own military capability (a decision that would backfire disastrously when the newly empowered domains turned their guns on the Shogunate itself).[55] It began purchasing steam-powered warships via the existing Dutch East India Company trading post at Nagasaki. And it started setting up military colonies at the outer limits of the realm in the hope of countering foreign territorial encroachment.[56] For centuries the Shogunate had banned ethnic Japanese from settling in its northern borderland, hoping to forestall confrontations with Indigenous groups and, later, the Russians. But in 1855 it reversed this policy and began encouraging domains to colonise the islands of Hokkaidō, Sakhalin and the Kuriles.[57]

52 'Summary of News from the Far East', The London & China Telegraph, 18 March 1861; Daniel V. Botsman, *Punishment and Power in the Making of Modern Japan* (Princeton: Princeton University Press, 2007), 133.
53 Cassel, *Grounds of Judgment*, 93.
54 D. Colin Jaundrill, *Samurai to Soldier: Remaking Military Service in Nineteenth-Century Japan* (New York: Columbia University Press, 2016).
55 Ravina, *To Stand with the Nations of the World*, 84. 56 *Ibid*, 92.
57 David L. Howell, *Capitalism from Within: Economy, Society, and the State in a Japanese Fishery* (Oakland: University of California Press), 61; Kikuchi Isao,

It was as part of this policy that the Shogunate began planning to colonise the Bonin Islands as well. The Shogunate was well aware of the existence of the Bonins. As early as 1670 some castaway fishermen had returned, reporting on the existence of two deserted islands far out to sea. Each was no more than forty kilometres in circumference, with a sweltering climate ('in summer it becomes so hot that you cannot walk barefoot on the rocks'), swarms of flies and mosquitoes, but also teeming with shellfish, sea bream and snapper, octopus, whales and a host of unfamiliar plants and fish.[58] The report so piqued the Shogunate's curiosity that it despatched an ocean-going galleon to survey the islands with an eye to possibly settling them, although nothing ever came of the venture.[59] Various other colonisation schemes were floated over the years, most notably by the scholar Hayashi Razan during a panic about Russian incursion in the 1780s. The vastly more perilous security environment of the mid-nineteenth century, however, changed the political calculus – especially once word got out that a small band of foreigners had established a colony on the islands already.[60]

In 1860, therefore, the Shogunate dispatched one of its new Dutch-built warships for the Bonins. The ship made landfall at Port Lloyd, the Bonins' main settlement, on 17 January 1862. Officials made preliminary inquiries, visiting islanders in their homes laden with an array of gifts: alcohol, textiles, snuff-boxes and toys for the children.[61] Nathaniel Savory was summoned to a tent erected on the beach within clear range of the *Kanrin Maru*'s guns. There Mizuno Tadanori, a court aristocrat and high-ranking Shogunal official, proceeded to explain to him that the Bonins had in fact been Japanese territory ever since their discovery by a samurai named Ogasawara Sadayori, and that the government had now returned to resume its administration.[62] He

Goryōkaku no tatakai: Ezochi no shūen (Tokyo: Yoshikawa Kōbunkan), 19–22.

58 Tanaka Hiroyuki, *Bakumatsu no Ogasawara: Ōbei no hogeisen de sakaeta ikan no shima* (Tokyo: Chūō Kōron, 1997), 5–6.

59 *Ibid*, 7.

60 Hyman Kublin, 'The Ogasawara Venture (1861-1863)', *Harvard Journal of Asiatic Studies* 14 (1951): 268; Tanaka, *Bakumatsu no Ogasawara*, 74, 157.

61 Kikuchi Sakujirō, *Bakumatsu Ogasawaratō nikki* (Tokyo: Ryokuchisha, 1983), 56–7.

62 Kublin, 'The Ogasawara Venture', 275. This story was almost certainly apocryphal, but circulated widely at the time and appears to have largely been taken at face value.

then issued a bilingual English-Japanese law code that protected the islanders' rights to the land they currently cultivated, but forbade them from cultivating new land without permission.

As an attempt at asserting territorial sovereignty, Mizuno's mission initially appeared to be a success. The islands' existing population generally agreed to abide by the law code, and to acknowledge that the islands belonged to Japan in a very material sense. In one telling exchange Savory insisted to Mizuno that the feral goats that roamed the islands belonged to him, for he had imported them from his home country. Mizuno retorted that the goats might be Savory's, but the grasses and trees they grazed on were Japanese. Savory seems to have accepted this argument, for he subsequently agreed to apply for permission before hunting them.[63]

More importantly, neither Britain nor the United States protested Japan's claim to administer the islands. British assent was tacit, but the American ambassador to Japan went so far as to explicitly recognise this 'reclamation of sovereignty' so long as Japan promised to 'protect' American merchants who were resident on the island.[64] British and American diplomats might even have supported the Japanese claim. To the extent that either government had a Bonin policy, it was to keep the islands open as ports for trade and to prevent them from falling into the sphere of influence of the other. On the other hand, neither Britain nor the US actually wanted to go to the trouble of administering the islands directly. In this sense, having a weak state such as Japan administer the Bonins was ideal – both as a cost-saving measure and to prevent the islands from falling into the hands of a strategic rival.[65]

Territorial sovereignty had its limits, however, and the Tokugawa colonisation project was quickly bedevilled by the problem of how to govern the islands while respecting the extraterritorial privileges claimed by some of its inhabitants. The presence of Hawaiians and other Pacific Islanders did not exercise Mizuno much, and indeed early reports dispatched from the island display a striking lack of interest in the 'kanaka' – a term that originally simply meant 'human being' in Hawaiian but in the mid-nineteenth century was used as a derogatory

63 Kikuchi, *Bakumatsu Ogasawaratō nikki*, 61.
64 OEC Ogasawara jūmin taiwasho: Taunsundo Harisu, 2 December 1861.
65 For a thorough discussion of this geopolitical calculus see Takahiro Yamamoto, 'Balance of Favour: The Emergence of Territorial Boundaries Around Japan, 1861–1875' (PhD diss., London School of Economics, 2015): 210.

term for Pacific Islanders more generally.[66] This reflected an assessment that dealing with 'kanaka' was unlikely to cause diplomatic headaches for the Tokugawa. Beachcombers, however, were another matter entirely – for any Bonin resident who could claim citizenship of an imperial power such as Great Britain, France, Spain or the United States was entitled to extraterritorial protection according to treaties that the Tokugawa had signed with those countries. For precisely this reason the detailed interviews recorded by Tokugawa officials reveal an exquisite concern to determine the nationality of beachcomber.

Complicating the question of extraterritorial jurisdiction further was the Shogunate's attempt to found a colony of permanent Japanese settlers on the Bonins. To recruit settlers Mizuno looked to another remote Pacific island, Hachijōjima, that lies roughly midway between mainland Japan and the Bonin Islands along the same oceanic ridge. For most of recorded history Hachijōjima was so remote as to be, in a literal sense, the edge of the known world.[67] Early modern Japanese cartographers often conflated it with the demon island of Nyōgashima, said to be populated by lascivious cannibal women who ensnared sailors through erotic reverie.[68] Actual life on Hachijōjima was probably not quite so much fun, for the island's annals are filled with tales of wrenching poverty and horrific famines. These were diagnosed by Japanese annalists as the result of chronic overpopulation, which seemed to recommend the island as a recruitment site for potential settlers.[69]

Even more promisingly, like other regions of Japan that ended up sending large numbers of people overseas, the island already had

66 Mizuno would later gloss 'kanaka' as 'races of the southern islands' (*nantō no jinshu*). See Katsu Kaishū, *Kaigun rekishi* 4 no. 10 (Tokyo: Kaigunshō, 1889), 9. During his initial survey of the island his delegation only bothered to record one interview with a 'kanaka' – a man named 'Jako' (ジャコ). Jako seems to have divided his time between sharecropping Savory's land in Okumura and working his own plot of land in Fukurozawa. But even then Mizuno did not bother to ask where Jako had come from, or how he had arrived on the islands. See OEC Ogasawara jūmin taiwasho; Katsu, *Kaigun rekishi* 5 no. 11: 24, 31–4.

67 Nelly Naumann, *Japanese Prehistory: The Material and Spiritual Culture of the Jōmon Period* (Wiesbaden: Harrassowitz Verlag, 2000), 54.

68 D. Max Moerman, 'Demonology and Eroticism: Islands of Women in the Japanese Buddhist Imagination', *Japanese Journal of Religious Studies* 36 no. 2 (2009): 357.

69 Kawasaki Fusagorō, *Edo jidai no Hachijōjima: kotō ku no kyūmei* (Tokyo: Tōkyō-to, 1964), 171.

a tradition of out-migration within the borders of Japan itself (Figure 1.3).[70] Since the seventeenth century the Shogunate had used Hachijōjima as a penal colony, and the annual ships that dumped prisoners on the island returned carrying second and third sons from islander families who set up branch villages on the mainland.[71] Hachijōjima also maintained a government office in Edo, primarily for the purpose of marketing the islands' much sought-after silk brocade. This meant that it was a relatively simple manner for the Shogunate to enlist a Hachijōjima official for the initial scouting expedition.

The official, Kikuchi Sakujirō, spent several days scaling Peel Island's steep hillsides (at one point nearly tumbling to his death), but then returned home persuaded that the Bonins had potential as a destination for out-migration. He summoned Hachijōjima's village headmen together, and presented them with a proposal.[72] The Shogunate would allocate the colonists as-yet-uncultivated land and a gift of five *ryō* each, and also provide food and housing 'for as long as it took for you to be able to weave your own clothes'.[73] He also tempted them with spiritual rewards: 'As the founding ancestors of the islands, you will be worshipped by your descendants as if you were *kami*'. Exiles were forbidden from enlisting, but poverty was not an obstacle. Colonists were required to be of good character and unmarried; skilled artisans such as carpenters and blacksmiths were also promised return passage to Hachijōjima at the end of fixed-length contracts if they desired.[74]

Kikuchi succeeded in recruiting a total of thirty-eight colonists, the majority of whom were landless.[75] Like the Hawaiians before them, they travelled to the Bonins bearing suites of crops, livestock, technologies and cultural practices to help them colonise the new islands. On arrival they set about establishing a new settlement in the next bay along from Port Lloyd, and started clearing, ploughing and seeding farmland with tools

[70] Martin Dusinberre, *Hard Times in the Hometown: A History of Community Survival in Modern Japan* (Honolulu: University of Hawai'i Press, 2012), Ch.6.

[71] Kawasaki, *Edo Jidai no Hachijōjima*, 208, 212–13.

[72] Kikuchi Sakujirō, *Bakumatsu Ogasawaratō nikki* (Tokyo: Ryokuchisha, 1983), 64.

[73] *Ibid,* 196, 215; Inamura Hiromoto, 'Ogasawara-jima ni okeru shiseki oyobi enkaku', *Rekishi Chiri* 48 (1926): 233.

[74] Kikuchi Sakujirō, *Bakumatsu Ogasawaratō nikki*, 196–7; Okuyama Ikuko, 'Hachijōjima ni okeru jinkō ryūshutsu katei to sono tokushitsu: omo ni Meiji shoki kara Dai-Ni Sekai Taisen made', *Chigaku Zasshi* 95 no. 1 (1986): 49.

[75] Ikuko, 'Hachijōjima ni okeru jinkō ryūshutsu': 49.

Figure 1.3 Hachijōjima Islanders gather to send off a ship (c.1845)
Source: Shigeyama Takanobu, *Izu Shichitō emaki* (1845). Columbia University C. V. Starr collection

provided by the Shogunate and cultivars brought from Hachijōjima.[76] This choice of site suggests that Mizuno initially planned for the Japanese settlers to form a distinct community from the Hawaiians and beachcombers. All the Tokugawa planning for the colony suggests a primarily terrestrial vision of settler colonialism aimed at attaining agricultural self-sufficiency. Indeed, if the Hachijōjimans had only stuck to farming their own plots of land after their arrival this might have made the job of administering the islands much easier.

This was not the case however, for the new arrivals soon found themselves drawn into the Bonins' maritime economy. This was in large part thanks to a man named Nakahama 'John' Manjirō, whose remarkable career had begun in 1841 when, as a fourteen-year-old fisherman, he had been swept out to sea and washed up on a desert island teeming with giant birds. He had then been rescued by an American sea-captain, studied English and navigation in Massachusetts, travelled the world crewing New England whalers, gotten rich in the San Francisco gold rush, and eventually returned to Japan, where he had been arrested for flouting the Shogunate's ban on overseas travel. He survived this ordeal and was later able to parlay his linguistic and intercultural fluency into an appointment as official translator aboard Mizuno's Bonin colonisation project.[77]

So taken was Nakahama by the abundant seas surrounding the islands that he decided to martial his prior experience and set up his own whaling operation. And his close government ties helped him persuade the Shogunate that development of the whale fishery was essential to ensuring the success of the Bonin colony. The Shogunate provided extensive state backing for his venture. It rented the boat, paid the crew's salary, and also provided Nakahama with employment as a translator during the off-season. In return, Nakahama agreed to hire a crew comprised mainly of Hachijōjiman colonists.[78] To train them he also hired six beachcombers who had prior experience in the New Bedford whaling fleet, and were thus familiar with latest technological

[76] Tanaka, *Bakumatsu no Ogasawara*, 187.
[77] For Nakahama's account of his overseas adventures see Ikaku Kawada, *Drifting Toward the Southeast: The Story of Five Japanese Castaways – a Complete Translation of Hyoson Kiryaku (a Brief Account of Drifting Toward the Southeast) as Told to the Court of Lord Yamauchi of Tosa in 1852 by John Manjiro* (New Bedford: Spinner Publications, 2003).
[78] Ishihara, *Kindai Nihon to Ogasawara-tō*, 204.

developments (such as the explosive harpoon) that were new even to Nakahama.[79] One of these was a US national by the name of William Smith who had recently absconded from a steamer en route from St. Petersburg to Nagasaki, then hidden out in the hills until it left port. Another, George Horton, had sailed aboard Perry's expedition but had jumped ship when the flotilla stopped off in the Bonins, and resided there ever since.[80]

It was these two men whom Nakahama accused of piracy after an exhausting session of whale-flensing. The dispute started local but ended up spiralling into a major diplomatic incident. Nakahama first returned to Port Lloyd, the Bonin Islands' main harbour, and turned Smith and Horton over to face trial. In doing so, he placed the Japanese magistrate there in a difficult position. The two men had been accused of a crime against a Japanese national, aboard a ship sailing under Japanese colours, in waters of islands that were supposed to be Japanese territory. On the other hand, Japan's treaty with the United States stated that any Americans resident on Japanese soil had the right to trial in consular courts. To complicate matters further, the same treaty also forbade foreigners from residing outside the limits of designated treaty ports. As Port Lloyd had not been designated a treaty port, Smith and Horton were technically in violation of Japanese law merely by being on the islands in the first place. Yet by 1862 Horton had lived in the Bonins for almost a decade, since before Japan began to administer the islands.

In the event, the Japanese magistrate punted. After consulting with Savory and other prominent Bonin beachcombers he despatched the two suspects to Yokohama aboard the *Ichiban Maru* to face trial in a consular court.[81] The ensuing trial partly vindicated Nakahama's accusation, but also exposed the way in which extraterritorial imperialism helped to defend the interests of beachcombers and, by doing so, compromised Japan's territorial sovereignty over the Bonin Islands. The court found Smith guilty, jailed him for four months, and then banned him from

[79] Many areas of mainland Japan, such as Wakayama, Kochi and Aomori, had long histories of coastal whaling. See Arch, *Bringing Whales Ashore*; Jonas Rüegg, 'Business, State and Environment in the Making of Japan's Kuroshio Frontier', in *Oceanic Japan*, ed. Stefan Hübner; Nadin Heé, Ian Jared Miller and William Tsutsui (Honolulu: University of Hawai'i Press, forthcoming).

[80] Tanaka, *Bakumatsu no Ogasawara*, 214–15. [81] *Ibid*, 219.

entering Japan. But it found in favour of Horton, demanding that he be returned to his landholdings on the Bonin Islands. When Tokugawa officials refused, the US consul fired off a blistering note, threatening to unleash 'the naval force of the United States' unless Horton was either returned to the Bonins or compensated with an indemnity of $32,000. Japanese officials eventually caved, and decided to pay the indemnity.[82]

From Borderland to Bordered Land

In 1863 the Shogunate abruptly decided to abandon the Bonin colonisation project. The Hachijōjimans packed up their belongings and returned home almost as swiftly as they had arrived. The exact reasons for this decision are still somewhat mysterious. One theory is that a new faction with different policy priorities came to power within the Shogunate. Another is that the Shogunate simply decided that any project that raised the likelihood of disputes between Japanese subjects and foreigner citizens was more trouble than it was worth. The full ramifications of the Horton affair only became clear *after* the decision to evacuate was taken, so could not have influenced it.[83] But they certainly illustrated the potential pitfalls of the project, and in later years Japanese officials would cite the risk of potential conflicts with foreigners as a reason against resuming colonisation.[84] Indeed, when the Japanese government did resume administration of the islands some twelve years later, the same issues soon re-emerged.

Much had changed in the intervening years. In 1868 the Tokugawa Shogunate collapsed, overthrown by a coalition of rebellious domains and imperial courtiers. Western diplomats, frustrated by what they saw as the Shogunate's intransigence and double-dealing, provided covert support for the rebels. A new government took power, ruling in the name of the young Meiji emperor. It embarked on an ambitious project to dramatically reshape the Japanese polity: centralising power, standardising legal codes and sweeping away the elaborate, decentralised

[82] Chapman, *The Bonin Islanders*. 64-5; Department of State, *Papers Relating to Foreign Affairs, Accompanying the Annual Message of the President to the Second Session Thirty-Eighth Congress* (Washington, DC: Government Printing Office, 1864), 67.

[83] For instance, the US consul did not issue his verdict on the Horton case until August 1863, two months after the Hachijōjiman colonists were evacuated. See Tanaka, *Bakumatsu no Ogasawara*, 208–9.

[84] Ishihara, *Kindai Nihon to Ogasawara-tō*, 225–6.

structure of the container society to create a direct relationship between the state and its citizens.[85]

As part of this self-consciously modernising project, the new Meiji government also embarked on an effort to transform the country into a fully sovereign state according to the emerging logic of territoriality. But it immediately confronted exactly the same diplomatic headaches that had plagued its predecessor. In 1871 a delegation of diplomats embarked on a world tour aimed at restoring tariff autonomy and abolishing foreign extraterritorial privileges, but returned in failure.[86] Meiji diplomats made more headway in demarcating Japan's borders with neighbouring states. But even here progress was halting. In 1872, just as the new government was abolishing the old feudal domains, it counter-intuitively established a new one by having the King of the Ryūkyū Islands swear an oath of vassalage to the Meiji emperor. This fudge was designed to avoid alienating China; the Ryūkyū kings had been sending tribute to both countries for centuries – a fact that both the Qing and the Tokugawa had chosen to ignore. Meiji diplomats worried that a bald declaration that the Ryūkyūs were Japanese territory would alienate its powerful neighbour and waited another nine years before finally deposing the king and establishing Okinawa Prefecture. Even after this, the southernmost border of the new prefecture was not delimited until well into the 1880s.[87]

Similarly, the new Meiji government achieved a diplomatic success in negotiating a territorial border between Japan and Russia. Tokugawa diplomats had waged a fruitless campaign to have Russia recognise Japan's sovereignty over the southern half of the island of Sakhalin, but had succeeded in only in establishing the principle of 'mixed residency'. In theory this had meant that the two countries would jointly administer the island as a Russo-Japanese condominium; in practice it had led to chaos.[88] The Meiji government resolved the matter by the simple expedient of ceding its claim to Sakhalin, in return for Russian recognition of Japan's sovereignty over the Kurile Islands. But even this success was at first a qualified one. Early Japanese attempts to colonise the group were frustrated by American sealers' tactic of using

[85] For a comprehensive discussion of the Tokugawa-Meiji transition in global context see Ravina, *To Stand with the Nations of the World*.
[86] Ravina, *To Stand with the Nations of the World*, 141–4.
[87] Bukh, *These Islands Are Ours*, 223.
[88] Cassel, *Grounds of Judgment*, 89; Ravina, *To Stand with the Nations of the World*, 92.

extraterritorial privileges to violate Japan's anti-poaching regulations with impunity. Eventually the government decided to evacuate its Kurile settlements entirely rather than risk provoking a diplomatic confrontation with the United States.[89]

The Meiji government's attempt to (re)claim the Bonins proceeded in a similarly chequered pattern. In many ways, beachcomber management represented an even more daunting challenge to Japanese rule than it had a generation previously. In 1862 only a handful of Bonin residents could claim beachcomber status, with the rest hailing from Hawai'i or other Pacific islands. But by 1876 many of those beachcombers had fathered children by Pacific Islander wives, with the result that a whole new generation of Bonin Islanders might possibly be able to claim beachcomber citizenship by paternal descent. A Japanese census conducted in March 1877 estimated, cautiously, that of the seventy-one people 'resident' on the island as many as fifty-seven could claim citizenship of either Britain, the US, France, Portugal or Spain – all countries that had signed extraterritorial treaties with Japan.[90]

The diplomatic uncertainty about the jurisdiction to which beachcombers and their families should be subject bedevilled the Meiji government's early efforts to administer the Bonin Islanders. In 1875 an expedition led by Obana Sakusuke, a veteran of the Tokugawa expedition, sailed for the Bonins to inform their inhabitants that Japanese administration had resumed. No sooner had it departed, however, than the British consul at Yokohama dispatched a gunboat hot on its heels. The official reason for the gunboat's visit was to investigate a complaint by a Briton called Webb, who had written to the consulate two years earlier asking for help in redressing a property dispute. But it was also a fairly unsubtle means of asserting extraterritorial sovereignty over those British citizens residing on the islands.

Webb's complaint was, in its labyrinthine complexity and sprawling spatial scope, fairly typical of the borderland Pacific. Millichamp, one of the original 1830 colonists, had cleared a plot of land known as Blossom Fields on the northern tip of Peel Island. In 1851 Millichamp had left the islands for the Spanish colony of Guam, 1,500 kilometres to the south, but the cadastral survey drawn up by Commodore Perry two

[89] Toshihiro Higuchi, 'The Nature of Unequal Treaties' paper delivered at the Association of Asian Studies annual conference, Washington, DC (24 March 2019).

[90] Obana Sakusuke, *Ogasawara Yōroku* 1 no. 19 (Tokyo: Yūunsha, 2014), 37.

years later nevertheless listed Blossom Fields as his property. At some point after this Webb had claimed the fields for his own, and succeeded in having the Tokugawa Shogunate's 1862 expedition recognise his ownership of the land in its own cadastral code. Subsequently Webb, too, left the Bonins to set up a sawmill on even more remote Ascension Island in the Carolines. But the sawmill venture ended in disaster, and when Webb returned to the Bonins he found Blossom Fields occupied by a Frenchman named Leseur who had visited Millichamp on Guam and bought the deed issued by Perry from him.[91]

These sort of squabbles were a common feature of life on the Bonins, but the dispute over Blossom Fields had an unusual significance. For one thing, by the time Webb returned from the Carolines Blossom Fields had become a particularly valuable piece of real estate, at least by Bonins standards: it now supported not only Leseur and his family of four but also two families of sharecroppers. Secondly, Webb was an important figure in Bonin Island society: he had first arrived in 1847, which made him a veritable old-timer among a population that was constantly in flux; he was the closest thing the Bonins had to a pastor, baptising the majority of children born on the islands;[92] and, perhaps most importantly, the death of Savory in 1874 left him as the only islander who could read and write, and therefore a vital line of communication with the outside world.[93] Indeed, it was this very literacy that enabled Webb to call for consular support in the first place. His request flew in the face of the logic of territorial sovereignty, for he was asking the British government to help him defend a title granted by the Japanese government against a rival title that was endorsed by an official of the United States. But in the context of the nineteenth-century Pacific it made a certain amount of sense, for as we have seen the British government regularly asserted the right to adjudicate disputes involving its subjects, even if they occurred within the territory of other countries.

The British consular aide, Russell Robertson, arrived at Port Lloyd two days after the Japanese expedition. He promptly summoned Obana's officials aboard the deck of his gunboat, and invited them to

[91] UKNA FCO 46/195: Robertson to Parkes (13 December 1875), 115.

[92] Chapman, *The Bonin Islanders*, 95.

[93] Obana, *Ogasawara Yōroku* 1 no. 19: 'Zaitō kaku kuni jinmei zen-koseki no gi o-todokesho' 37; Russell Robertson, 'The Bonin Islands', *Transactions of the Asiatic Society of Japan* 9: 20 October 1875–12 July 1876 (Tokyo: Hakubunsha, 1888), 138.

watch as he interviewed Webb about the details of his complaint. He then suggested that the two governments launch a joint investigation into the Webb–Leseur dispute. The officials gave a non-committal reply, quite correctly perceiving it as an attempt to undermine Japan's territorial sovereignty over the Bonins. This refusal to engage created something of an impasse, whereby the Japanese expedition then watched warily as Robertson and his men traipsed around the island, conducting enquiries and generally making the presence of Her Majesty's Government felt.[94]

This caution eventually paid off. After completing his investigation, Robertson ultimately declined to issue a ruling on the Webb–Leseur dispute, and after two weeks returned to his ship and made sail back for Yokohama. There the matter lay until six months later Webb, frustrated by the lack of British consular assistance, eventually turned to Japanese officials for help instead. But Obana tartly rejected his suit, ruling that Leseur should be entitled to Blossom Fields as he had been the one cultivating it when Japanese officials had returned to the island the previous year.[95]

That British diplomats did not ultimately challenge Obana's ruling was in one sense a diplomatic victory for Japan. Unlike the dispute between Nakahama, Smith and Horton, the Webb–Leseur dispute did not end up being tried in a consular court. Instead, it was resolved through the Japanese justice system. Still, this did not mean that Anglo governments had yielded the principle of extraterritoriality entirely, for heated diplomatic discussions continued over what it might mean to subject Anglo nationals in the Bonins to Japanese law. The British ambassador continued to insist to the Japanese foreign minister that British nationals resident in the Bonins be subject to the trial through the consular courts.[96] And the US ambassador noted with alarm that proposed custom regulations for the Bonins would violate Japan's treaty obligations 'in so far as they subject citizens of the United States to trial and judgement in Japanese tribunals'.[97]

In response the Japanese government decided to resolve the issue by taking a different tack, pressuring islanders to relinquish their citizenship entirely. In 1876, Obana ordered all Bonin residents to enter their

[94] UKNA FCO 46/195: Robertson to Parkes (13 December 1875), 127.
[95] Obana, *Ogasawara Yōroku* 2 no. 109: 'Ei no Uebu Butsu Rezuwaa Susaki-mura jisho shusso no ketsumatsu', 122–3.
[96] Ishihara, *Kindai Nihon to Ogasawara-tō*, 258. [97] *Ibid*, 86–7.

names into the government's new nationwide family register, thereby renouncing their existing citizenships and 'naturalising' as subjects of the Japanese emperor. To convince the islanders he offered a mix of threats and blandishments. Naturalisation would be rewarded with gifts of clothing, blankets, tobacco and alcohol, and money to buy farming implements, furniture and building materials for housing. Refusal, however, would leave the islanders bereft of protection from Japanese law.[98] But Robertson got wind of this plan and fired off a memorandum reminding the islanders that 'by becoming Japanese subjects they would place themselves under Japanese laws and would ... forfeit all rights to British protection'. This was a coded warning to the islanders, but also to the Japanese government as well.[99]

As a result, only five islanders obeyed the naturalisation order. Of these, three were Pacific Islanders who had jumped ship from an American frigate nine years previously. Within what passed for Bonin society they had relatively low status, all living as subordinate members of households headed by beachcombers. And all were in Japanese eyes viewed simply as 'kanaka' prior to their naturalisation. Two of them, Kopepe and Friday Tomoto, originally hailed from the Marquesas, an island group that at the time was under only the most nebulous form of colonial 'protection' by France. Whatever their affective sense of belonging they might reasonably have felt that in renouncing their Marquesan nationality they were not giving up something of much practical diplomatic value. The third 'kanaka' to naturalise, Sam Tempory, came from a mysterious island recorded as Tamantsu (タマンツ島). Only two years after becoming a Japanese citizen he joined the crew of a passing whaler, never to return.[100]

The other two applicants for Japanese citizenship were colonial subjects of European powers: a Bermuda-born Briton and a Manila-born Spaniard. Both men had arrived in the Bonins only a few years previously and had married Japanese wives who were themselves recent arrivals. They, too, were relatively marginal members of Bonin society, both spatially and economically, living essentially as share-croppers renting plots of land from Leseur at Blossom Field. The Briton, Myers, even wrote to the British consulate explaining that he

98 *Ibid*, 259–60. 99 Chapman, *The Bonin Islanders*, 90.

100 Obana Sakusuke, *Ogasawara Yōroku* 4 no.107 (Tokyo: Yūunsha, 2014): 'Kikajin Samu Tenporii joseki taitō no ken', 81–2.

feared being murdered by another islander and hoped that Japanese citizenship might offer him a modicum of protection.[101] Tellingly, the Japanese government handled their cases with much more delicacy, despatching diplomatic notes to inform the British and Spanish embassies of their naturalisation.[102]

Still, even this effort was only partially successful. The British consul retorted that until the Japanese civil code had an internationally recognised naturalisation law on the books, it would not even be theoretically possible for a British citizens to become Japanese. This was a particularly shrewd riposte for it exploited the truth that the definition of Japanese citizenship was indeed still in flux, as the Meiji government struggled to knit together a welter of village and domain-level censuses into a national household register system.[103] Of course Japan was hardly alone in this respect, for British citizenship was not exactly a clearly demarcated category either: confusion over whether or not children of mixed Anglo-Indian parentage were British nationals persisted until the end of the British Raj, for instance.[104] But this did not matter. The realities of nineteenth-century imperialism meant that Britain reserved the right to decide the limits of the Japanese body politic, if not in a territorial sense then certainly in a juridical one.

The End of Beachcomber Sovereignty

In 1882, six years after Obana had first ordered them to, the remaining Bonin Islanders agreed to enter their names in Japan's household registry, implicitly forsaking their right to extraterritorial protection from their countries of birth. This date marks the point at which Japan gained, to all intents and purposes, full territorial sovereignty over the Bonin Islands. In one sense, it represents the triumph of sustained diplomatic, juridical and economic efforts to transform the archipelago

101 Obana Sakusuke, *Ogasawara Yōroku* 2 no.52 (Tokyo: Yuūnsha, 2014): 'Kikajin Marisu Ei ryōjikan to sho ōfuku shokan utsushi', 49.

102 Obana, *Ogasawara Yōroku* 1 no.18: 'Zaitōjin nyūseki no gi honshō e ukagaesho', 34.

103 Chapman, *The Bonin Islanders*, 90–1. In the event, Japan did not enact a naturalisation law until 1899. On the origins of the Meiji Japanese household register system, see Endō Masataka, *Koseki to kokuseki no kingendaishi: minzoku, kettō, Nihonjin* (Tokyo: Akashi Shoten, 2014).

104 Valerie E. R. Anderson, 'The Eurasian Problem in Nineteenth-Century India', (PhD diss. The School of Oriental and African Studies, 2011), 37.

into a predominantly Japanese settlement. These efforts included resuming the effort to transplant colonists from Hachijōjima, so that by the time the islanders agreed to naturalise they were already outnumbered by new arrivals more than four to one.[105] It also involved placing restrictions on immigration from outside Japan, a measure aimed at preventing new beachcombers from settling on the islands.

But other factors were at work, too. For one thing, by the 1870s the Bonins had conspicuously failed to live up to their promise as a maritime provisioning hub. The shift to steam power meant that more and more ships could cross the Pacific in one leg without needing to stop for provisions. Traffic from whalers had diminished also, partly as a result of overhunting but also because lamp oil manufacturers were struggling to compete with cheaper, better-smelling kerosene.[106] Surviving logbooks from American whalers show a sharp drop-off in whale sightings near the Bonins between 1850 and 1875, and almost no sightings after 1875.[107] As a result the number of ships that made port at the islands declined also. Between January 1833 and July 1835 some twenty ships had called at Port Lloyd, and Savory claimed that at one point even thirty or forty per year had not been unusual. By the 1870s the average was barely one a year. Fewer ships meant fewer opportunities to barter turtle meat for matches, linens or grog, or to sign on for stints crewing whalers. With the islands increasingly isolated from the broader trans-Pacific extractive economy, life on the Bonins increasingly became a matter of eking out a living through agriculture. Given this, it is not obvious that many beachcombers would have been drawn to live on them even if they had been permitted to.[108]

Other factors external to Japanese government policy may also have helped bring an end to beachcomber sovereignty as well. While no direct testimony survives to shed light on why the islanders chose to naturalise in 1882, the immediate catalyst seems to have been the death of Webb. It was Webb who had led the resistance against

105 Ishihara, *Kindai Nihon to Ogasawara-tō*, 268.

106 David, Gallman and Gleiter, *In Pursuit of Leviathan*, 148; Demuth, *Floating Coast*, 61; Arch, *Bringing Whales Ashore*, 29–32, 184.

107 Smith et al., 'Spatial and Seasonal Distributions of American Whaling and Whales in the Age of Sail', 14.

108 Chapman, *The Bonin Islanders*. 30; Conversation between Nathaniel Savory and Mizuno Tadanori, 26 January 1862. OK vol.8. cited in Yamamoto, 'Balance of Favour', 202–3; UKNA FCO 46/195 Robertson to Parkes (23 December 1875).

naturalisation, in large part thanks to his existing relationship with Robertson, and through him with the British consulate at Yokohama, but also because he was the only one of the islanders who could read and write.[109] Bereft of literacy, the remaining Bonin Islanders lost their only possible means of communicating with the British or American diplomatic staff who might come to their aid.

The islanders may also have felt that, even if they had been able to make contact with Anglo embassies, there was little chance that help would be forthcoming. After the deaths of Savory and Webb, the vast majority of the population were second-generation Bonin Islanders, and under both British and US law could claim nationality only by proving they had been born within wedlock to a parent born in the respective country. Yet the official account of Perry's expedition did not regard 'children born on the islands', even those born to American fathers, as 'native Americans'.[110] With the notable exception of Thomas Webb's family, British consular officials regarded the relationships between Hawaiians and beachcombers with distaste. Whereas Japanese officials recorded female partners as *tsuma* (meaning partner in a legal sense), Robertson preferred 'mistress'.[111] And he reported that Nathaniel Savory's Chamorro widow:

> had been living as the companion of two other men at different times prior to her becoming a companion of Savory. It is questionable therefore how far [she and her six children] . . . may be entitled to American protection, and any other claims to American [or British] citizenship by the settlers would be of the same shadowy nature.[112]

Remarks such as these whiff of mingled prurience and racism, but casting aspersions on the parentage of islanders also served a practical purpose. As we have seen, imperial states often viewed beachcombers as something of a nuisance. As citizens they had symbolic importance: prestige was on the line, and letting them be treated roughly by a foreign government might set an alarming precedent for the future. And it is true that consular officials who dealt with beachcombers directly often genuinely felt aggrieved on

109 UKNA FCO 46/195 NA FCO 46/195: Robertson to Parkes. (27 December 1875).
110 Hawks, *Expedition of an American Squadron*, 203.
111 Obana Sakusuke, *Ogasawara Yōroku* 1 no.19: 37–9; UKNA FCO 46/195: Robertson to Parkes (23 December 1875).
112 UKNA FCO 46/195: Robertson to Parkes (23 December 1875).

their behalf, especially when they felt authentic ties of ethnic kinship. But beachcombers also had a habit of getting themselves into trouble in remote corners of the world, creating a host of diplomatic headaches in the process. The foreign ministers who reviewed their cases from a distance tended to look on them much more coolly. In some cases gunboats were dispatched to intervene directly, but where possible imperial states preferred to use weaker states as intermediaries, negotiating extraterritorial treaties and setting up systems of consular courts.

When it came to the children of beachcombers, an even simpler solution presented itself: disavow citizenship completely. The erosion of beachcomber sovereignty on the Bonins can be understood not just as the result of assertive Japanese colonial policy, but as a story of imperial abandonment.[113]

Borderland Citizens

As newly minted Japanese subjects, the descendants of the original Bonin colonists found themselves in an awkward position, for their status as 'naturalised foreigners'[114] also fell short of full citizenship. They were denied the vote, banned from residing elsewhere in Japan, and were generally regarded with a blend of suspicion, contempt and ethnographic curiosity by their new government.[115] Themselves the progeny of settler colonists, they soon found themselves outnumbered in the islands by an influx of new, government-subsidised Hachijōjiman settlers. Those Hachijōjimans also, it should be said, enjoyed only a qualified form of citizenship. Though the Meiji government allowed Japanese on the mainland to elect their own local government officials, it did not extend this privilege to the archipelago's remote islands. Instead, with unmistakeable condescension, it despatched bureaucrats to rule islands such as the Bonin, Izu and Amami Islands directly 'on account of the environment, the nature of the people, and their old-fashioned customs'.[116] In this sense we could even speak of Hachijōjima as an 'internal colony' in the same mould as Hokkaidō or Okinawa, and of Hachijōjiman settlers in the Bonins as

113 See also Yamamoto, 'Balance of Favour', 210. 114 *kika gaikokujin.*
115 Chapman, *The Bonin Islanders*, 93.
116 Takaesu Masaya, *Kindai Nihon no chihō tōchi to tōsho* (Yumani Shobō, 2009), 20, 44.

subaltern colonists in the same mould as Koreans, Taiwanese, Scots and Corsicans.[117]

At the same time, Bonin and Hachijōjima residents enjoyed some privileges denied to their mainland compatriots. Although Port Lloyd was never formally designated an international trade port, in practice the local authorities continued to allow foreign ships to call there. And though the Bonins never regained their status as a whaling hub, they did attract traffic from hunters in pursuit of other oceanic prey. This meant that well into the twentieth century Bonin Islanders found it relatively easy to sign up for stints on Kurile-bound sealing expeditions, at a time when ordinary Japanese had to apply for exit permits to leave the country.[118] Like their beachcomber and Hawaiian forbearers, not all returned.

Seals were not the only marine animals that beckoned. Even after the colonisation of the Bonin Islands many other North Pacific islands remained uninhabited, their resources as yet untapped. Bonin and Hachijōjima residents, by dint of their location and oceangoing experience, were ideally positioned to profit from those resources. To do so they did a number of things. They staged elaborate performances to style themselves as pioneers capable of transforming wilderness into permanently habitable space. They leveraged their status as Japanese citizens to lobby their own government for support, in the process drawing new connections between private claims to property and state claims to territory. And they slaughtered untold millions of seabirds.

117 Mason and Lee, *Reading Colonial Japan*, 18–20, Barbara Brooks, 'Peopling the Japanese Empire: The Koreans in Manchuria and the Rhetoric of Inclusion', in *Japan's Competing Modernities: Issues in Culture and Democracy, 1900–1930*, ed. Sharon Minichiello (Honolulu: University of Hawai'i Press, 1998); Seiji Shirane, *Gateway Imperialism: Colonial Taiwan and Japanese Expansion in South China and Southeast Asia, 1895–1945* (Ithaca: Cornell University Press, 2022); Linda Colley, *Britons: Forging the Nation 1707–1837* (New Haven: Yale University Press, 1992), 127–30; Robert Aldrich, 'France's Colonial Island: Corsica and the Empire', *French History and Civilization* (August 2017): 112–25.

118 Ishihara, *Kindai Nihon to Ogasawara-tō*, 287; H. J. Snow, *In Forbidden Seas: Recollections of Sea Otter-Hunting in the Kurils* (London: Edward Arnold, 1912) 151–3. For the history of Japanese passports see Kamishiraishi Minoru, 'Meiji Ishin ki ryoken seido no kisoteki kenkyū', *Shien* 73 no. 1 (2013); Takahiro Yamamoto, 'Japan's Passport System and the Opening of Borders, 1866–1878', *Historical Journal* 60 no. 4 (2017): 997–1021.

2 The Race to Marcus Island

Commodities and Colonisation in the North Pacific, 1885–1902

In the summer of 1902 two ships raced from opposite sides of the Pacific towards a reef-ringed speck of rock covered only in birds and birdshit. The rock was named Marcus Island, or Minami-torishima in Japanese, and it was tiny. Even at its highest point it poked barely twenty metres above sea level, and its coral sand beaches could be strolled comfortably in under an hour.[1] Yet despite its minuscule size Marcus now threatened to provoke a diplomatic confrontation between the United States and Japan, two empires that in the previous years had expanded across the Pacific towards each other with startling rapidity.

The race was triggered on 10 July, 1902, when a Honolulu-based sea-captain named Captain Andrew Rosehill, recently invested stock-holder in the Marcus Guano Island Co., announced his intention to sail to the island aboard the sail-ship *Julia E. Whalen* 'armed with credentials from the States Department of the United States, which show that he has been granted title to the land by America'.[2] In Washington, DC, the Japanese ambassador read the news of the *Whalen's* departure with alarm. He immediately telegraphed his superiors in Tokyo, pointing out that the island had been incorporated within the Ogasawara (Bonin) Islands municipality four years previously, and advising the dispatch of a man-of-war to protect Japanese interests there.[3] While Japanese diplomats considered how to respond, the press got hold of the story: the *Tokyo Asahi Shimbun* warned of 'US plans to occupy our isolated island', and the *Yomiuri* followed up the next day with

1 Shoichi F. Sakagami, 'An Ecological Perspective of Marcus Island, With Special Reference to Land Animals', *Pacific Science* 15 no. 1 (1961): 82; Kokudo Kōtsū-shō, 'Minami-torishima no gaiyō' (Tokyo, [n.d.]): www.mlit.go.jp/common/001025368.pdf. Last accessed on 13 June 2020.

2 'Want Cable at Marcus Island', *The Hawaiian Star*, 10 July 1902.

3 MOFA 1.4.1.7 Teikoku hanto kankei zakken: Takahira to Tokyo, 13 July 1902.

similarly alarmist coverage.[4] On 23 July, nearly two weeks after the *Whalen* had left Honolulu, Tokyo telegraphed Washington explaining that Marcus Island was already Japanese sovereign territory. That same day the fast cruiser *Kasagi* departed for Marcus with a Japanese diplomat onboard, so as to impress upon Rosehill the reality of Japanese jurisdiction over the island.

Immediately upon hearing news of the *Kasagi*'s departure, however, the Marcus Island Guano Co.'s backers revealed to the Honolulu press what they had previously seen fit to conceal: anticipating resistance from Japanese 'squatters' on the island, they had loaded the *Julia E. Whalen* with Mauser repeating rifles.[5] The stage was set for conflict, and with both ships ocean-bound (and thus cut off from all means of communication), the media were left to speculate.[6] Which ship would reach Marcus first? The island was a full three days' sail west of Honolulu, and only three days' east of Japan. But the *Whalen* had a substantial lead. If the *Whalen* arrived first, would they evict the Japanese by force of arms? If the *Kasagi* succeeded in repelling the *Whalen* on its arrival (as seemed likely given the large disparity in firepower), how would the American media react to this? Would the US government feel compelled to respond in turn by sending its own warship? On tiny Marcus Island, two expanding great powers in the Pacific seemed about to collide.

Chapter 1 used the colonisation of the Bonin Islands to explore how two global phenomena, the age of extraction and extraterritorial imperialism, created a juridical borderland in the North Pacific. This chapter uses the Marcus Island Incident of 1902 as a case study through which to continue the story into the twentieth century. It shows how, even after the collapse of the whale fishery, demand for new commodities such as guano and plumage triggered a speculative scramble to control remote Pacific islands and the seabird colonies that nested on them. But it also shows how this scramble played out within a changed diplomatic context. The age of extraterritorial empire in the North

4 'Minami-torishima to Beikoku seifu', *Tokyo Asahi Shimbun*, 27 July 1902; 'Beikoku kojima senryō keikaku ni tsuite', *Yomiuri Shimbun*, 17 July 1902.

5 'Where Were the Mausers?' *The Hawaiian Star*, 31 July 1902; 'Mausers for Men on Marcus: Secret Orders for Captain of Whalen', *The Hawaiian Gazette*, 1 August 1902.

6 'Rondon tokuden', *Tokyo Asahi Shimbun* (16 July 1902); MOFA 1.4.1.12 Kokka oyobi ryōiki ni kan suru zakken: 'Is Our Flag Over Marcus Island?'

Pacific had given way to one in which governments increasingly drew on new theories of international law to assert territorial sovereignty over insular oceanic space. These new theories grounded sovereignty in a particular human relationship to nature, by which the claimant asserted his right to property and territory by colonising it and thereby 'improving' it from its natural state. To maintain such claims, guano miners and bird-hunters had to stage performances of colonisation on islands that often proved frustratingly resistant to habitation.

The Guano Rush

On the cusp of the twentieth century the age of extraction continued apace, much of it continuing to target the marine and terraqueous creatures of the North Pacific. Walruses were rendered into umbrella handles, and beavers, seals and sea otters were transformed into an array of luxuriant coats and stoles.[7] Much of this extraction was rampantly unsustainable, but the depletion of the very resources on which the hunters relied only prompted efforts to search for them farther afield. Of these, two resources in particular trained human attention on Marcus, an island long known to navigators but until then paid little heed.[8]

The first of these was guano, or calcified bird excrement. Guano accumulates over hundreds or thousands of years, on and around the nesting sites of seabirds such as pelicans, cormorants and albatrosses.[9] Its use as fertiliser has a very long history indeed. Control of the commodity played an outsized role in the political economy of the Inca Empire (the term guano itself derives from the Quechua word '*huanu*' meaning dung), and the accounts of European travellers along the Pacific coastline of South America mention active guano mines from as early as the seventeenth century.[10] But the scale of mining underwent a dramatic step-change in the 1840s after merchants discovered the stupendous

[7] McNeill, 'Of Rats and Men'; Bathsheba Demuth, 'The Walrus and the Bureaucrat: Energy, Ecology, and Making the State in the Russian and American Arctic, 1870–1950', *The American Historical Review* 124 no. 2 (April 2019): 483–510.

[8] Bernhard Welsch, 'Was Marcus Island Discovered by Bernardo De La Torre in 1543?', *The Journal of Pacific History* 39 no. 1 (1 June 2004).

[9] Cushman, *Guano and the Opening of the Pacific World*, 169–70.

[10] *Ibid*, 3.

profits to be had exporting guano to Britain, Europe and the US.[11] Between 1846 and 1850, guano exports to Britain alone increased tenfold, to nearly 200,000 tons annually.[12] A year later, an American diplomat estimated that the Peruvian guano trade was worth about $3 million per year.[13]

This new global guano trade involved immense brutality. The actual work of mining involved long stretches of physical labour under the hot sun breathing the acrid stench of guano dust. Compounding this was the fact the guano miners were mostly recruited from China and were effectively debt peons: to pay for the voyage across the Pacific they had either signed long-term indentured contracts or had borrowed money from family or local moneylenders in their home villages. With no way to return home even if they wished to, they were exposed to the whims of the overseers. Suicides occurred regularly.[14] Guano mining also decimated bird populations, for it involved stripping away the rock directly underneath nesting sites. It was not until the twentieth century that development economists seriously considered the possibility that guano mining might be conducted on any kind of sustainable basis.[15]

At the same time, the guano trade yielded astonishing profits for investors, many of whom accrued vast fortunes. William Gibbs, who in 1847 won the exclusive contract to export Peruvian guano to Britain and Europe, earned up to £100,000 per year during the 1850s (about £13,000,000 in today's money) which he used to build a Neo-Gothic country pile in Somerset.[16] In the words of one memorable ditty that circulated around the City of London: 'The House of Gibbs made their dibs, Selling the turds of foreign birds.'[17] The scramble for control of

[11] *Ibid*, 25–6.

[12] Jimmy M. Skaggs, *The Great Guano Rush: Entrepreneurs and American Overseas Expansion* (London: Palgrave Macmillan, 1994), 9–10.

[13] Skaggs, *Great Guano Rush*, 14.

[14] Edward D. Melillo, 'The First Green Revolution: Debt Peonage and the Making of the Nitrogen Fertilizer Trade, 1840–1930', *American Historical Review* 117 no.4 (October 2012): 1039.

[15] Cushman, *Guano and the Opening of the Pacific World*, 149.

[16] Skaggs, *Great Guano Rush*, 7; Matthew Kilburn, 'Gibbs, William (1790–1875)' in *Oxford Dictionary of National Biography* (19 May 2011); Bank of England Inflation Calculator: www.bankofengland.co.uk/monetary-policy/inflation/infl ation-calculator. Last accessed 29 November 2019.

[17] David Kynaston, *The City of London: A World of Its Own, 1815–1890* (London: Chatto & Windus, 1994): 171.

the resources even triggered a war between Spain and Peru, Chile, Ecuador and Bolivia for control of the guano-rich Chincha Islands.[18]

Lured by the promise of immense fortunes, prospectors began scouring the oceans in search of guano islands to claim.[19] The initial phase of the 'guano rush' focused primarily on the Southeast Pacific, as the unusually dry climate there helps to preserve the nitrate content of bird faeces. But it gained a new lease of life when farmers also began to use guano for its phosphate content – the second of the three macronutrients identified by Justus von Liebig as a limiting factor in plant growth.[20] Phosphatic guano did not yield the same astounding profits as nitrate guano but it was found more widely, from the Caribbean to the North Pacific to the South China Sea. Several promising deposits were identified within striking distance of Hawai'i: Baker and Palmyra were worked between the 1850s and 1870s, and another mine opened on the even more remote island of Laysan in 1890 (Figure 2.1).[21]

Sovereignty, Territory, Colonisation

It was against the background of the ongoing guano rush that one Andrew Rosehill alighted on the atoll of Marcus Island in 1889. There was much of the beachcomber about Rosehill. He was born in Sweden in 1852 but by his early twenties was working as a seaman based out of San Francisco, from where he naturalised as a US citizen.[22] Shortly after this he relocated to Hawai'i, where he married a Native Hawaiian woman and became active in local politics. In the 1890s, for instance, he served in the Honolulu police force and then in the Citizens' Guard, a militia set up to defend the annexationist Republican government (that had itself overthrown the Hawaiian

[18] Brett Clark and John Bellamy Foster, 'Ecological Imperialism and the Global Metabolic Rift: Unequal Exchange and the Guano/Nitrates Trade', *International Journal of Comparative Sociology* 50(2009): 324–8.

[19] Skaggs, *Great Guano Rush*, Ch.2.

[20] Clark and Foster, 'Ecological Imperialism and the Global Metabolic Rift', 316, 330.

[21] Christina Duffy Burnett, 'The Edges of Empire and the Limits of Sovereignty: American Guano Islands', *American Quarterly* 57 no. 3 (September 2005); Gregory Rosenthal, 'Life and Labor in a Seabird Colony: Hawaiian Guano Workers, 1857–70', *Environmental History* 17 (October 2012).

[22] California State Library: Great Registers, 1866–1898. Accessed via ancestry.ca on 11 January 2022.

(a)

(b)

Figure 2.1a and 2.1b I-Kiribati guano miners and albatrosses on Laysan Island (c.1891)

Source: L. Walter Rothschild, *Avifauna of Laysan* (1893)

monarchy in a US-backed coup).[23] And although he made Hawai'i his home, as a sea captain by trade he also spent much of his time criss-crossing the wider Pacific. In fact, his first visit to Marcus came during a stint spent trading in Micronesia, during which he had conceived the notion of developing the atoll as a coconut plantation.[24]

Rosehill's report to the US State Department describes Marcus as an island 'about five miles in length, by about two miles in width', reefed with coral and buffeted by winds to the extent that 'anchorage ... is available for only about four and one-half months in the year'. As a result there were, he claimed, 'no signs of human occupancy, savage or civilised.' Indeed, he insisted, he was certain that he was 'the first human being to set foot upon its soil'. Yet formidable as Marcus was to access, the island was home to millions of seabirds, its surrounding waters were teeming with fish, and after digging into its soil to a depth of several feet Rosehill estimated it contained over 100,000 tons of guano.[25]

Given the potential wealth to be had from mining guano, Rosehill's interest in Marcus was understandable. But how was he to go about claiming its guano for himself? The rush to claim guano islands played out within a larger global scramble to demarcate the world into territorial units that could be administered through new, centralised institutions of government.[26] As part of this effort, international legal theorists attempted to hammer out some kind of diplomatic protocol over how to establish sovereignty over uninhabited territory. It appeared first in the writings of jurists such as Vattel and Wheaton, then found its way into international treaties, and eventually into institutions such as the Hague Permanent Court of Arbitration,

23 'Kapena Makai hou ko Honolulu', *Ka Nupepa Kuokoa* (2 February 1895); 'A Big New Squad', *The Hawaiian Star* (22 February 1895). In 1894 Rosehill was granted 'the privilege of Special Citizenship' for his efforts protecting the fledgling Republican government, although this did not entail renouncing his US citizenship. See 'Certificate of Special Rights of Citizenship' (23 July 1894). Accessed via ancestry.ca on 11 January 2022.

24 NARA-II RG 59 State Dept Guano Islands, 'In the Matter of the Application of Andrew A. Rosehill for the right to remove Guano deposits from Marcus Island' (20 January 1902).

25 *Ibid.*

26 Martti Koskenniemi, *The Gentle Civilizer of Nations: The Rise and Fall of International Law, 1870–1960* (Cambridge: Cambridge University Press, 2001), 116–17.

established in 1899 specifically to adjudicate sovereignty disputes between states.[27] And it was based on two principles.

The first was that title to uninhabited space could only be asserted through 'peaceable, continuous occupation'. This concept drew on aspects of property law dating to medieval European thought, but twisted during the early modern period to argue that occupation (of property or, later, territory) ought to involve not only habitation but also 'improvement'.[28] In the late nineteenth century international jurists therefore came to understand sovereignty as possessing an explicit temporal aspect: the credibility of a claim rested partly on the extent to which the claimant could show they were developing their territory from a deserted state to a cultivated one. It embraced, in other words, what Richard Drayton calls 'the economics of Genesis . . . the idea that the command of resources was made legitimate by their most efficient use'.[29]

The second principle, which began to gain major purchase during the Berlin Conference (1884–1885), was that sovereignty over a territory was contingent on administration by a state. This was a much newer innovation, arising from a series of scandals over 'rogue empires' set up by private adventurers in Borneo and later the Congo.[30] The principle of state administration was important because it had the potential to invert the relationship between sovereignty and property rights. In the past, prospectors had claimed islands as their personal property, and states had then used these property claims as a foundation upon which to assert sovereignty. Under the new principle, though, territorial administration ought to precede property claims, because it was held that only the territorial state could assign property rights.

Both these legal concepts contained considerable scope for ambiguity, and indeed were often deployed with alarming cynicism. In Africa

[27] Lauren Benton and Benjamin Straumann, 'Acquiring Empire By Law: From Roman Doctrine to Early Modern European Practice', *Law and History Review* 28 no. 1 (2010), 26; Koskenniemi, *Gentle Civilizer of Nations*, 213; Susan Pedersen, *The Guardians: The League of Nations and the Crisis of Empire* (Oxford: Oxford University Press, 2015), 172.

[28] Andrew Fitzmaurice, *Sovereignty, Property, and Empire, 1500–2000* (Cambridge: Cambridge University Press, 2017), 2–3.

[29] Richard Drayton, *Nature's Government: Science, Imperial Britain, and the 'Improvement' of the World* (New Haven: Yale University Press, 2000), 229.

[30] Steven Press, *Rogue Empires: Contracts and Conmen in Europe's Scramble for Africa* (Cambridge, MA: Harvard University Press, 2017), 242.

and Australia, European colonisers were wont to expropriate Indigenous land on the grounds that slash-and-burn agriculture or nomadic pastoralism did not constitute 'effective occupation' – heedless of the fact that nomadic pastoralists and fishermen often developed common-property regimes for husbanding resources.[31] Conversely, claims of occupation were often mere fig-leafs – the mere chartering of a company being considered sufficient to claim sovereignty over vast tracts of unmapped interior.[32] This was particularly the case for bird islands, whose harsh environments and sheer remoteness made them arduous to settle and cumbersome to administer. Still, by the late nineteenth century both concepts – of 'continuous occupation' and 'state administration' – were circulating sufficiently widely that claimants to marginally habitable territory were at least vaguely aware that they might prove decisive in resolving potential sovereignty disputes.

Anatomy of a Failed Claim: The Spratly Islands, 1877–1932

We can see how these two principles helped to shape sovereignty claims by examining a short-lived guano-mining operation on a group of atolls and cays known as the Spratly Islands. Unlike Marcus Island and the Bonins, which sit atop seamounts jutting up from the deep ocean, the Spratlys lie within the South China Sea, a shallow basin of water hemmed in by land and long traversed by fishermen and merchant junks. During World War II they were annexed by Japan (see Chapter 4), and at the time of writing they are the object of a complex and bitter sovereignty dispute involving China, the Philippines, Taiwan, Indonesia, Vietnam and Brunei. Before any of these states launched formal claims to the islands, however, an Anglo-American consortium of guano prospectors set out to mine them under the flag of Britain. The first record of this venture dates to 1877, when a C. R. Graham and two partners submitted a claim to two of the islands in the

[31] Koskenniemi, *Gentle Civilizer of Nations*, 108, 115; James Fairhead and Melissa Leach, *Misreading the African Landscape: Society and Ecology in a Forest-Savanna Mosaic* (Cambridge: Cambridge University Press, 1996); Elinor Ostrom, *Governing the Commons: The Evolution of Institutions for Collective Action* (Cambridge: Cambridge University Press, 1990).

[32] Koskenniemi, *Gentle Civilizer of Nations*, 123–4.

British colony of Labuan, off the coast of Borneo.[33] William Treacher, acting governor of Labuan at the time, provisionally registered the claim subject to approval of the Foreign Office. He then forwarded the claim to London, but it quickly became mired in an interdepartmental back and forth. The Foreign Office was inclined to back Graham's claim, being reasonably sure that no foreign government had already claimed sovereignty over the two islands. But the Colonial Office argued that the islands fell beyond the limits of Labuan, and that any claim to them should therefore be registered with Britain's Consul General in Borneo.[34]

Two years went by without any resolution to this question, until an impatient Graham decided to begin mining the islands anyway. He hired thirty Chinese labourers from Hong Kong, shipped them to the islands, and set them to work mining guano. The venture soon ended in disaster. Within weeks some of the men began asking to return to Hong Kong, a request that Graham rejected as a breach of their indenture contract. Eventually a dispute over food rations ended up with Graham's English assistant shooting two miners dead. The entire party arrived in Labuan, where Graham accused nineteen of his labourers of attempted murder. Treacher, ignoring the Colonial Office's explicit instruction that the Labuan government should have no jurisdiction over the Spratly Islands, convened a jury to try the case anyway. Two of the miners were found guilty and sentenced to two years of imprisonment with hard labour.[35]

For those homesick, hungry Chinese miners (especially the two who ended up doing time on a chain-gang), British jurisdiction no doubt felt very real. And the administrative confusion did not stop the Foreign

33 UKNA T 161/622: Straits Settlement Government Gazette, 27 November 1879, Government Notification No. 432; Owen Rutter, *British North Borneo: An Account of its History, Resources and Native Tribes* (London, Bombay, Sydney: Constable & Co., 1922), ix. As Graham told it, they chose to register their claim in Labuan because it was the closest British territory to the two islands in question. But they may also have sought to take advantage of a moment of political flux in the region, when British rule was on the verge of expanding beyond the small entrepôt at Labuan to encompass the larger territorial swathe of North Borneo.

34 UKNA T 161/622: Foreign Office to Attorney General and Solicitor General, 16 March 1932 (2095/178/17).

35 UKNA T 161/622: Treacher to Foreign Secretary, 30 October 1879; 'Labuan: From the Daily Times, November 11th', *Straits Times Overland Journal* (14 November 1879), 4.

Office placing the Spratlys on a list of miscellaneous islands 'that are British territory, or under British protection, but are not included into any Colony or separate Protectorate.'[36] In this sense the South China Sea of the 1870s constituted a juridical borderland not unlike the Bonin Islands at around the same time.

Yet in later years British diplomats were not so sure whether their claim to the islands would hold up in a court of international law. When in 1932 the French government annexed the Spratlys Britain chose not to contest the claim. For one thing Graham's Spratly Island mine had shuttered its doors soon after the debacle of 1879, and later efforts to mine guano had also come to nothing. But Foreign Office diplomats were just as worried by the administrative ambiguity of the situation. One confessed that Treacher's 'action in trying these offenders may have been misconceived', as the islands had never been officially placed under Labuan's administration. But in that case, which part of the British Empire did administer them? The official concluded that 'an adverse decision [by an international court of arbitration] might well draw attention to other instances at which H. M. G. would be equally hard put to prove effective annexation.'[37]

US Bird Island Policy, 1850–1902

Britain was not the only government that found itself wrangling over how to define the sovereign status of bird islands. In 1850, a decade into the guano rush, the US government found itself at loggerheads with Spain over the island of Contoy off the coast of Honduras. Two years later a diplomatic scandal over the Lobos Islands off the coast of Peru forced Secretary of State Daniel Webster from office.[38] Next, in 1856, the US became embroiled in a three-way dispute with the British and Venezuelan governments over the guano island of Aves in the Caribbean.

[36] UKNA T 161/622: Foreign Office to Attorney General and Solicitor General, 16 March 1932 (2095/178/17); *Straits Settlement Government Gazette*, Notification No. 438 21 November 1879.

[37] UKNA T 161/622: Foreign Office to the Law Officers of the Crown; Treasury Secretary to Treasury Solicitor, 24 August 1932: 'Spratly & Amboyna Islands. Annexation by the French Republic'.

[38] Skaggs, *Great Guano Rush*, 22–6.

In the wake of this, Congress began to mull the potential implications of the guano rush for American foreign policy. On the one hand, Virginia tobacco planters clamoured for guano to fertilise their crops.[39] On the other, allowing every which prospector to claim bird islands as US territory could make the United States liable to defend, in perpetuity, sovereignty over a collection of remote islets whose only conceivable value lay in the finite amount of guano that could be mined from them.

In an attempt to forestall this eventuality, Senator Seward (who would later, as Secretary of State, negotiate the purchase of Alaska) proposed a piece of legislation that became known as the Guano Islands Acts, designed to delineate sovereignty over 'only these more ragged rocks, which are covered with this deposit in the ocean, which are fit for no dominion, or for anything else, except for the guano which is found on them'. The Guano Islands Acts were juridically innovative in that it formulated a new form of sovereignty, temporally limited depending on the material conditions of the islands themselves. Namely, 'whenever the Guano should be exhausted, or cease to be found on the islands, they should revert and relapse out of the jurisdiction of the United States'.[40]

However, the legislation was dogged by a lack of clarity. It did not specify how to determine that guano deposits on a given island were exhausted. The cessation of mining activity was not necessarily a guide, for guano exhaustion was only one possible reason for the closure of a mine. Mining operations tended to commence, halt and resume depending upon a number of factors including operating cost, the projected market price of guano and (as we saw on the Spratlys) the shifting political economy of labour. There was not even any certainty as to whether or not guano constituted a renewable resource – it being unclear the rate at which bird excrement was deposited and calcified.[41] In some cases the State Department requested claimants to purchase a bond from the US Treasury, which could be redeemed only when the claimant declared that the guano deposits had been

[39] Steven Stoll, *Larding the Lean Earth: Soil and Society in 19th-Century America* (New York: Farrar, Straus and Giroux, 2003), 194.

[40] Cong. Globe, 34th Cong. 1st sess. (23 July 1856): 1697–1700, cited in Burnett, 'Edges of Empire', 785–6.

[41] George Hutchinson, 'The Biogeochemistry of Vertebrate Excretion', *Bulletin of American Natural History* 96 (1950).

exhausted. But this rule was only enforced intermittently, and in fact the State Department issued a wide range of sometimes contradictory responses to applications.[42]

Neither did the Guano Island Acts provide much clarity as to how exactly bird islands were to be administered. As Christina Ponsa-Kraus argues, the islands' constitutional status as 'appertaining' to the United States was clearly an administrative fudge. The result was that the implementation of the Acts resembled nothing so much as 'bureaucratic chaos'. State Department officials later concluded that 'it carried no exact meaning and lends itself readily to circumstances and the wishes of those using it'.[43] For this reason US diplomats hesitated to invoke it in any subsequent disputes over bird islands, such as with Britain and Hawai'i over atolls such as Baker and Palmyra Island.[44] By the same token, at the time of writing the US government has also never declared sovereignty over any specific guano island to have expired.[45]

We can surmise Rosehill's own understanding of the relationship between sovereignty, property and territory from his subsequent attempts to claim Marcus Island as his own. Before departing the island he managed to convince a crew-member and his I-Kiribati wife to decamp on the island, where they set about planting garden seeds, raising pigs and chickens, and building a 'substantial native house'.[46] Rosehill himself set sail bearing a sample of the island's guano, which he submitted for chemical analysis back in Honolulu. He was satisfied enough with the sample's commercial value that he then wrote to the State Department, requesting title to the island under the Guano Islands Acts.[47] He received a dubious reply, asking him to provide proof that Marcus had not in fact been claimed by another government, and also a bond as surety against his promise to mine the island.

Such proof never arrived, however, and beyond this point Rosehill's efforts petered out. Perhaps he failed to secure financing for his venture, or perhaps he was distracted by other schemes. We know, for instance,

42 Burnett, 'Edges of Empire', 787. 43 *Ibid,* 786–7.

44 *Ibid,* 780; Cushman, *Guano and the Opening of the Pacific World*, 104; Rosenthal, 'Life and Labor in a Seabird Colony'.

45 Burnett, 'Edges of Empire', 781.

46 NARA-II RG 59 Letters Received Relating to Guano Islands, 1852–1905: 'In the Matter of the Application of Andrew A. Rosehill for the Right to Remove Guano Deposits from Marcus Island' (20 January 1902).

47 NARA-II RG 59 Letters Received Relating to Guano Islands, 1852–1905: George Martin to Anon. (20 October 1890).

that in 1890 he was hired as the foreman on another guano mine on the island of Laysan (see Chapter 3).[48] By 1895 he was at the helm of another schooner, dividing his time between trading with Micronesia and fishing. The year after that he told *The Hawaiian Star* that he had claimed yet another atoll, this time in the vicinity of Hawai'i's French Frigate Shoals.[49] As for the couple Rosehill left on Marcus to begin colonising it, after spending eleven months on the island they grew weary of life there and boarded a passing vessel bound for Honolulu.[50]

The Bird Rush

During his absence, Marcus became enveloped in yet another 'commodity frontier', driven by what Hiraoka Akitoshi has aptly called the 'bird rush'.[51] There is nothing particularly new about an appreciation for feathers; indeed, they have been in demand as items of adornment for thousands of years.[52] The Hawaiian nobility (ali'i) were spectacularly voracious consumers of feathers, having a penchant for 'magnificent cloaks, woven of a mass of bright but tiny feathers into something like a variegated velvet . . .' To secure these totems, which were believed to protect the wearer from harm, ali'i would orchestrate systematic culls of bird colonies on islands such as Nihoa using 'a system of corvée labour, on . . . a Mayan or Egyptian scale'.[53]

In the late nineteenth century, however, feathers suddenly went from being an elite status symbol to a mass-market consumer product.[54] The trend began in Paris, the centre of the nineteenth-century fashion world. By 1876 *Le Moniteur de la Mode*, whose writers styled themselves as global tastemakers, was grumbling that women both young and old, urban and rural, had taken to promenading in bonnets

48 HSA M-476 Pacific Chemical Fertilizer Co: Minutes (20 August 1890).
49 'Cruise of the Norma', *The Hawaiian Star* (31 August 1896).
50 MOFA 1.4.1.12 Kokka oyobi ryōiki ni kan suru zakken: 'Marcus Island – The United States' Claims'.
51 Hiraoka, *Ahōdori to 'teikoku' Nihon no kakudai*, 110–11.
52 Robin W. Doughty, *Feather Fashions and Bird Preservation: A Study in Nature Protection* (Oakland: University of California Press, 1974), 6.
53 Thomas, *Islanders*, 74–6; Adrienne L. Kaeppler, 'Hawaiian Featherwork in the Age of Exploration', *in Royal Hawaiian Featherwork: Na Hulu Ali'i*, ed. Leah Caldeira, (Honolulu: University of Hawai'i Press, 2015), 41.
54 Sarah Abrevaya Stein, *Plumes: Ostrich Feathers, Jews, and a Lost World of Global Commerce* (New Haven: Yale University Press, 2008), 20.

Figure 2.2 Women's millinery fashion as illustrated in *Le Moniteur de la Mode* (1876)

festooned with down frills, ostrich feathers and sometimes even whole birds stuffed and mounted (Figure 2.2).[55] Soon after, magazines such as *Vogue*, *Harper's Bazar*, *McCalls* and *Good Housekeeping* began promoting plumed millinery in Britain and the United States, so that five years later fashion houses in New York were debuting bonnets ornamented with 'half a dozen humming-birds placed in a row, to show the brilliant hues of their breasts and throats'.[56] The fashion proved remarkably enduring. In 1897 the Paris correspondent for *Harper's Bazar* was still exclaiming about the latest French trend for hats made of 'rose-colored crêpe trimmed with Amazon plumes of the same fascinating color'.[57] Much as Hawaiian ali'i had once ventured into

55 Mary d'Auberville, 'Modes: Nouveautés, Description de Toilette', *Le Moniteur de la Mode* 44 (4 October 1876): 518.

56 Doughty, *Feather Fashions and Bird Preservation*, 14–15; 'New York Fashions: Bonnets and Hats', *Harper's Bazar* 14 (15 October 1881): 659.

57 Katherine de Forest, 'Our Paris Letter', *Harper's Bazar* 30 (4 December 1897): 1003.

battle armoured in plumage, so did fashionistas now sally into the salons of the *fin de siécle* (Figure 2.2).

This rise in demand for feathers inspired bird-hunters to fan out across the globe in search of previously untapped bird populations that might be converted into commodities for profit.[58] In the words of one fashion writer, 'that there should be an owl or an ostrich left with a single feather apiece hardly seems possible. It is certainly a great pity that a charming and perfectly feasible fashion should so soon be exaggerated into a craze.'[59]

The craze did indeed devastate global bird populations. Most famously, it led directly to the extinction of the North American passenger pigeon, a species that would become a sort of *memento mori* for wildlife conservationists. These birds had once been systematically hunted by Native Americans, but the pathogens that accompanied European colonisers decimated the Indigenous population and allowed pigeons to breed unchecked.[60] By the nineteenth century passenger pigeons were roaming across the continent in flocks large enough to blot out the sun for hours at a time. Then from the 1870s the species' population plummeted from an estimated 2 billion to extinction by the 1900s, mainly due to netting by commercial hunters supplying the plumage trade.[61]

Countless other species vanished across the world during the same period, and many more were driven to the brink of extinction or extirpated from entire habitats.[62] In densely populated Japan, bird populations never attained the kind of unchecked growth seen in North America, so the effect was less dramatic. But the raised price of plumage, combined with revised Meiji game laws that undermined Tokugawa regimes of common-property resource management, still decimated birdlife on the main islands.[63] By the 1890s the Meiji

58 Doughty, *Feather Fashions and Bird Preservation*, 61. For Japan, see *Umō to negu no hanashi: sono rekishi to bunka*, ed. Umō Kenkyūkai (Tokyo: Nihon Keizai Hyōronsha, 1993), 136–46.

59 'New York Fashions', *Harper's Bazar* 30 (4 December 1897): 1003.

60 Charles C. Mann, *1491: New Revelations of the Americas Before Columbus* (New York: Knopf, 2005), 318.

61 Joel Greenberg, *A Feathered River Across the Sky: The Passenger Pigeon's Flight to Extinction* (New York: Bloomsbury USA, 2014), xii.

62 Doughty, *Feather Fashions and Bird Preservation*, 87–8; Mark Barrow, *Nature's Ghosts: Confronting Extinction From the Age of Jefferson to the Age of Ecology* (Chicago: University of Chicago Press, 2009), 82.

63 Toshihiro Higuchi, 'Birds for New Japan: Bird Conservation and Reform, 1934-1952' delivered at the Association for Asian Studies Annual Conference (27 March 2010).

government was forced to implement a raft of new game laws that placed certain species off-limits and required hunters to secure permits to use either guns or nets.[64]

In the global market for plumage, longer feathers plucked from larger birds fetched a markedly higher price. In fact, length commanded such a premium that entrepreneurs in South Africa even set up ostrich farms dedicated primarily to feather production, a solution that proved ecologically but not economically sustainable. Others found it simpler to track down wild birds to kill. Rimbaud, having forsaken poetry for a career as a merchant in the Horn of Africa, wrote an 1883 report on the remote Ogaden region in which he noted excitedly that at a spot on the banks of the River Webi, 'hundreds of dollars' worth of ostrich feathers' could be collected 'in a few weeks'.[65]

As well as ostriches, albatrosses became particularly prized by plumage hunters, and it was primarily the search for this most quintessentially oceanic of megafauna that drove the bird rush out to sea. A distinctive interlocking shoulder bone mechanism enables albatrosses to glide on ocean thermals for weeks and even months on end, allowing them to hunt in deep-sea fishing grounds that are inaccessible to other birds. Unique access to such rich supplies of squid, fish and crustacea has allowed albatrosses to evolve to an enormous size. Royal and wandering albatrosses have an average wingspan of some three metres, while the albatrosses of the North Pacific typically grow to over two metres.[66]

Unfortunately, those same evolutionary traits that adapted albatrosses to the North Pacific made them uniquely valuable to, and uniquely vulnerable to, bird-hunters. Millions of years nesting on remote islands had left them bereft of any evolutionary memory of land-borne predators, and even after they learned to fear humans their large size rendered them slow and ungainly while on the ground, for launching themselves into the air required a long, lumbering takeoff of several meters. Small bands of bird-hunters could thus cull large

64 Rinyachō, *Chōjū gyōsei no ayumi* (Tokyo: Rinya Kōsai-kai, 1979), 5–6.

65 Stein, *Plumes*; Graham Robb, *Rimbaud* (London: Pan Macmillan, 2001), 348. Stein describes how ostrich farmers in South Africa were plunged into bankruptcy when demand for plumage dropped off after the outbreak of World War I.

66 Bernhard Grzimek, *Grzimek's Animal Life Encyclopedia Vol. VII: Birds* (New York: van Nostrand Reinhold, 1975), 141; Michael Brooke, *Albatrosses and Petrels Across the World* (New York: Oxford University Press, 2004), 7, 126.

numbers of albatrosses in a few hours' work. Japanese hunters would prove particularly adept at this, scouring the still largely uninhabited Bonin Islands for unsuspecting albatrosses. The journalist Hattori Tōru, observing a band of Japanese bird-hunters at work, described their technique as follows:

> To catch the birds, they approach them in parties of four to prevent the bird from flying up. They can only run with outstretched wings until they come to a slope or get a favourable gust of wind; so they are chased upward from below. Thus the birds in the reeds have to be surrounded, but the incubating birds are very easily approached. They are killed by striking them on the head with a club, and it is not difficult for a man to kill 100 to 200 birds daily.[67]

Boosters and Bird-Hunters

As with the guano rush, the bird rush inspired prospectors to set about claiming uninhabited islands for themselves, in order to monopolise access to the source of a highly valued commodity. In Japan the first person to do this was a man named Tamaoki Han'emon. Tamaoki hailed from a relatively affluent family of cattle-ranchers on Hachijōjima.[68] In 1862 he had participated in the Tokugawa Shogunate's abortive colonisation of the Bonin Islands, and in 1876 he had returned to set up shop as a sundries merchant. In these voyages he could not help but notice the increasing shipments of feathers to the mainland, as hunters stripped the islands of their bird populations to sell to foreign merchants in Yokohama.[69] And in his own voyages he would have regularly sailed past a deserted island, midway between Hachijōjima and the Bonin Islands, known simply as 'Torishima' (bird island) after the flocks of seabirds that teemed above it.

Until the 1880s Torishima (or Izu-Torishima) was known, if at all, simply as an island on which castaways washed up occasionally. It was there, for instance, that Nakahama Manjirō had been marooned before being rescued by an American whaler.[70] It was a place you got stuck on,

[67] Hattori Tōru, 'Bakadori no hanashi', *Dōbutsugaku Zasshi* 12 (15 October 1889): 411.
[68] Izu Shotō Tōkyō Ikan 100-nen Shi Hensan Iinkai, *Izu Shotō Tōkyō ikan 100-nen shi* (Tokyo: Tōkyō-to Tōsho Chōson-kai, 1981), 1067–71.
[69] Hiraoka, *Ahōdori to 'teikoku' Nihon no kakudai*, 20–1.
[70] Tanigawa Ken'ichi, *Hyōryū to hyōchaku sōsakuin* (Tokyo: Shōgakukan, 1993), 111.

not somewhere anyone might actually want to go to. It was also unusually challenging to access, even for a bird island, for it was ringed by steep cliffs with no obvious harbour. But Tamaoki was a man of impressive political connections and a talent for publicity, which enabled him to mobilise government support for his attempt to colonise the island.

To do so, Tamaoki successfully tapped into a global culture of imperialism that was sweeping the Pacific at the time. Between 1874 and 1900 Britain alone annexed Brunei, North Borneo, Sarawak, Fiji, Tuvalu, the Gilbert Islands (Gilbertese: Kiribati) the Kingdom of Tonga, and the island of Banaba with its rich phosphate reserves.[71] Spain, long an imperial presence in the Pacific, acquired the Carolines in 1885, and the same year ceded the Marshalls to the rapidly expanding German Empire.[72] Germany also claimed the Marianas, Nauru (another phosphate island) and a part of New Guinea, which became known as the 'Bismarck Archipelago'.[73] During the same period Germany began to jostle with Britain and the US over control of the Samoan archipelago as well, in a dispute that excited public passions out of all proportion to the islands' economic or strategic value.[74] One diplomat complained that 'even though the great majority of our pot-house politicians did not know whether Samoa was a fish or a fowl or a foreign queen, they shouted all the more loudly that whatever else it was, it was German and must remain forever German.'[75]

Japan, too, was consumed by imperial fever during this period. Even before the country's dramatic victories against China (in 1895) and Russia (in 1905), a loose coalition of politicians, intellectuals and entrepreneurs had been making the case for Japanese expansion into the *Nan'yō* (lit: South Seas) – a rather hazily defined region extending

71 Tony Ballantyne and Antoinette Burton, *Empires and the Reach of the Global, 1870–1945* (Cambridge, MA: Harvard University Press, 2012), 6. For the history of Banaba, see Katerina Martina Teaiwa, *Consuming Phosphate Island: Stories of People and Phosphate from Banaba* (Bloomington: Indiana University Press, 2014).

72 Mark R. Peattie, *Nan'yō: The Rise and Fall of the Japanese in Micronesia, 1885–1945* (Honolulu, University of Hawai'i Press, 1992), 5, 8. For an introduction to Spain's empire in the Pacific see Oskar Spate, *The Spanish Lake* (Canberra: Australia National University Press, 1979).

73 Ballantyne and Burton, *Empires and the Reach of the Global*, 7.

74 Pedersen, *Guardians*, 172.

75 Margaret MacMillan, 'The BBC Reith Lectures Episode 3: Civilians and War' (10 July 2018). Accessed via BBC iPlayer on 8 November 2020.

from Okinawa as far Hawai'i or even Australia.[76] The Imperial Japanese Navy supported the idea in the hope of securing increased budget allocations, and another key supporter was the Meiji oligarch Enomoto Takeaki, who as early as 1876 had hatched a plan to resettle a group of rebellious samurai on Micronesian islands he hoped to purchase from Spain.[77] Enomoto lent his support to various exploratory expeditions to the Pacific, and collaborated with the economist Taguchi Ukichi to found a government-subsidised trading company that would harvest marine products around the Bonins and undertake trade with Spain's Pacific colonies.[78] And the Mayor of Tokyo, Takamine Goroku, subsidised a quarterly shipping route that would sail to the Bonins in the hopes of spurring economic development of the islands.[79]

Unlike earlier, Tokugawa-era colonisation projects, the *Nan'yō* boom was highly publicised, for whereas the Shogunate had generally viewed the Japanese masses with suspicion Meiji imperialists hoped to enlist them. Enomoto established the Nan'yō Association, as well as the Tokyo Geographical Society (modelled after the Royal Geographical Society in London), to help collect and disseminate information about the region.[80] The navy also helped to whip up public enthusiasm for colonisation by embedding public intellectuals like Shiga Shigetaka and Hattori Tōru on its ships as they toured the Pacific.[81] These intellectuals in turn supplied Meiji Japanese readers with tales of adventure on the high seas, laced with pervasive angst about the trans-national vogue for empire. Shiga, probably the most influential *Nan'yō* booster of all, invoked ubiquitous theories of Social Darwinism when he warned that Japan must seize its own colonies if it was not to become colonised itself.[82] And Sugiura Shigetaka, who campaigned consistently for

[76] Peattie, *Nan'yō*, 2.
[77] J. Charles Schencking, 'The Imperial Japanese Navy and the Constructed Consciousness of a South Seas Destiny, 1872–1921', *Modern Asian Studies* 33 no. 4 (1999): 774; J. Charles Schencking, *Making Waves: Politics, Propaganda, and the Emergence of the Imperial Japanese Navy, 1868-1922* (Stanford: Stanford University Press, 2005), 44. Peattie, *Nan'yō*, 6.
[78] Peattie, *Nan'yō*, 16–18.
[79] TMA 617.B4.08: Meiji 21-nen Ogasawara shorui (kaikei-ka) no.14.
[80] Peattie, *Nan'yō*, 7. [81] Schencking, 'The Imperial Japanese Navy', 774.
[82] Shiga Shigetaka, *Nan'yō jijō* (Tokyo: Maruzen, 1887), 7, cited in Tierney Robert, *Tropics of Savagery: The Culture of Japanese Empire in Comparative Frame* (Oakland: University of California Press, 2010), 23.

Nan'yō colonialism throughout a career that spanned education, journalism and politics, lamented in an 1887 newspaper editorial 'how miserably small Japan looks on a map of the world'.[83]

The schemes promoted by *Nan'yō* boosters took a variety of forms, and in fact for many of its advocates the various kinds of colonial/imperial projects often tended to blur together into vague dreams of frontier-style expansion. Still, it is possible to tease out certain recurring strands. Some writers, such as the economist Taguchi, envisaged Japan carving out an empire of free trade on the British model (see Chapter 1). Others imagined subjugating indigenous populations and setting them to work on plantations. In 1892 the adventurer Suzuki Tsunemori returned from a voyage round the Pacific urging that Japan invade Hawai'i before it was snatched by a neighbouring power.[84] Others still imagined uninhabited islands that Japanese colonists might settle for themselves, and it is this strand of *Nan'yō* thinking that most directly concerns us here. For instance, in 1890 Shiga Shigetaka wrote that:

> Every year on the anniversary of Emperor Jimmu's accession ... we should ceremonially increase the territory of the Japanese empire even if only it is by a small measure. On each of these days our navy vessels should sail to a still unclaimed island, occupy it, and hoist the Japanese flag ... Not only would such a program have direct value as practical experience for our navy, but it would excite an expeditionary spirit in the demoralized Japan race.[85]

Japan's Bird Island Policy, 1887–1902

Tamaoki expertly tapped into this spirit of colonial boosterism in order to secure access to Torishima's albatrosses. Through contacts in Tokyo he heard that Enomoto, now serving as communications minister, had lent the explorer Yokō Tosaku a lighthouse patrol ship, the *Meiji Maru*, in order to survey the uninhabited Volcano Archipelago (J: Kazan Rettō). Tamaoki managed to wangle a position for himself aboard the expedition, and had Enomoto personally order that he be dropped off on

[83] Jun Uchida, 'From Island Nation to Oceanic Empire: A Vision of Japanese Expansion from the Periphery', 62, *Journal of Japanese Studies* 42 no. 1 (Winter 2016).

[84] Peattie, *Nan'yō*, 14.

[85] Shiga, *Nihon-jin*, 3 April 1890, 12, cited in Schencking, 'The Imperial Japanese Navy', 776–7.

Torishima en route.[86] This time Tamaoki succeeded in making landfall by paddling a canoe from the main ship into the islands' only harbour. As the *Meiji Maru* continued southwards, he and his crew proceeded to clamber to and fro across Torishima's steep cliffs. Within days they had found drinking water and erected a crude shelter, and set about slicing the wings off large numbers of seabirds, to be boxed and shipped back to the mainland.[87]

At this point, the expedition went awry. When Yokō's ship passed Torishima on its return journey the weather was again too rough to attempt a landing, and Tamaoki's crew was left marooned.[88] Subsequent events reveal Tamaoki's gift for tapping into the mood of *Nan'yō* excitement that swept up so many elite political figures in the 1880s. When the *Meiji Maru* returned to Tokyo, the Japanese press criticised the Tokyo government for abandoning the distressed pioneers. Even Tamaoki's own wife and daughters joined the effort, publicly pleading for the Postal Ministry to send a boat to rescue him.[89] Within a week another ship was dispatched to collect the stranded men.[90] What might have been construed as a debacle was instead conjured into heroism and eventually – when Tamaoki and his crew were rescued safely – as a triumph.

Ten days after his return to Tokyo, Tamaoki filed a petition for Torishima to be incorporated as Japanese sovereign territory under the jurisdiction of Tokyo Prefecture. By this point he was not the only person petitioning for access to Torishima. On 21 February Saitō Rinzō, the manager of a marine products trading firm, applied to the Tokyo prefectural government for a permit to fish on the island. But Tamaoki's bid prevailed, and in 1888 the Home Ministry granted him a free ten-year lease to the island. Just as crucially, the Tokyo government directed the Bonin ferry service to add Torishima to its route, supplying a crucial lifeline that would connect the island to the mainland.[91]

86 Mochizuki Masahiko, 'Tamaoki Han'emon to Torishima kaitaku', *Nantō Shigaku* 40 (October 1992): 45.
87 TMA 617 B4 8: Tamaoki Han'emon, 'Torishima taizai nikki'.
88 Mochizuki, 'Tamaoki Han'emon to Torishima kaitaku', 44; Hiraoka, *Ahōdori to 'teikoku' Nihon*, 22–3.
89 'Tamaoki no saijo', *Yomiuri Shimbun* (7 December 1887).
90 Hiraoka, *Ahōdori to 'teikoku' Nihon*, 23–4.
91 Mochizuki Masahiko, 'Tamaoki Han'emon to Torishima kaitaku', 45.

Significantly, Tamaoki's application took pains to emphasise that his venture aligned with the Japanese government's project to colonise uninhabited territory 'in both chilly Hokkaidō and the balmy Bonins'. His albatross, he promised, were only the first stage in a plan to develop the island in a way that would allow for permanent human settlement by promoting industries as varied as fishing, farming and cattle-ranching. In this way, Tamaoki claimed, leasing the island to him would help 'expand Japan's maritime possessions' in perpetuity by establishing a basis for continuous occupation.[92] His application to renew the lease ten years later likewise made a great show of his progress towards this end, and contained an elaborate map detailing his various infrastructure projects on the island, including port facilities, houses and a comprehensive road network (Figure 2.3).[93]

In the years that followed the global market price for feathers soared, as did Japan's total feather exports (Figure 2.4).[94] Tamaoki's bird-hunting activities made him a multimillionaire in today's terms and his son gained a reputation as a nouveau riche playboy who lived a life of luxury, yachting and driving to school in a motorcar.[95] The human settlement on Torishima also flourished, though not nearly so lavishly. By 1902 there were 125 people residing in Tamaoki Village year-round. The village even had its own school.[96]

The prosperity was gained at the direct expense of Torishima's bird population. Early visitors to Torishima described how nesting

[92] TMA 617 A5 10 'Torishima tokō-hi haishaku-gan': Tamaoki Han'emon, 'Torishima haishaku narabi teiki-sen go-kitō gan' (27 September 1887).

[93] TMA 625 D4 19 'Torishima ikkatsu': Tamaoki Han'emon, 'Torishima haishaku keizoku go-gan' (5 October 1897).

[94] Tōyō Keizai Shimpōsha, ed. *Nihon bōeki yōran* (Tokyo: Tōyō Keizai Shimpōsha, 1975), 30–1, 274; Great Britain Board of Trade, *Statistical Abstract for the United Kingdom in Each of the Last Fifteen Years from 1879 to 1893: No. 41* (London: H.M. Stationery Office, 1894) [C-7525].

[95] On his death in 1910 Tamaoki was reported to be worth as much ¥1 million, which would equal roughly $10 million dollars in today's money. The recollection about the younger Tamaoki's lavish lifestyle comes from the ornithologist Uchida Seinosuke (see Chapter 6), who happened to attend the same middle school. See 'Obituary', *The Japan Times* (3 November 1910); Uchida Seinosuke, *Tori: watashi no shizen-shi* (Tokyo: Sanseidō, 1971), 41. For price indices see the Bank of Japan website: www.boj.or.jp/announcements/education/oshiete/history/j12.htm/. Last accessed 9 November 2021.

[96] Kishōchō Torishima Kurabu 'Torishima' Henshū Iinkai, *Torishima* (Tokyo: Tōe, 1967); 'Torishima funka tansaku' *Tokyo Asahi Shimbun* (2 September 1902).

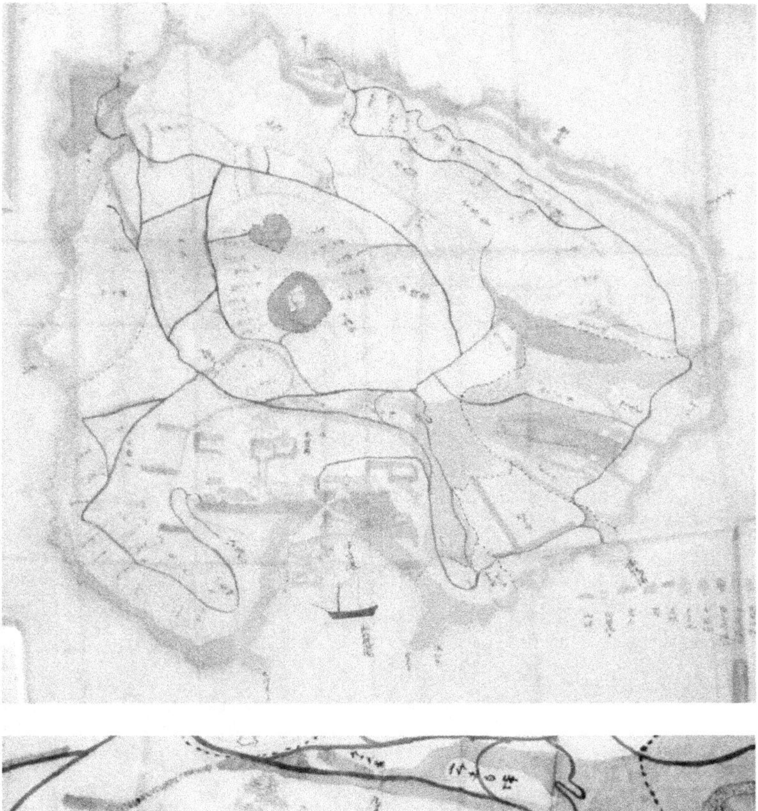

Figure 2.3 Map prepared by Tamaoki Han'emon showing his development of Torishima, 1888–1897
Source: Tokyo Metropolitan Archive collection

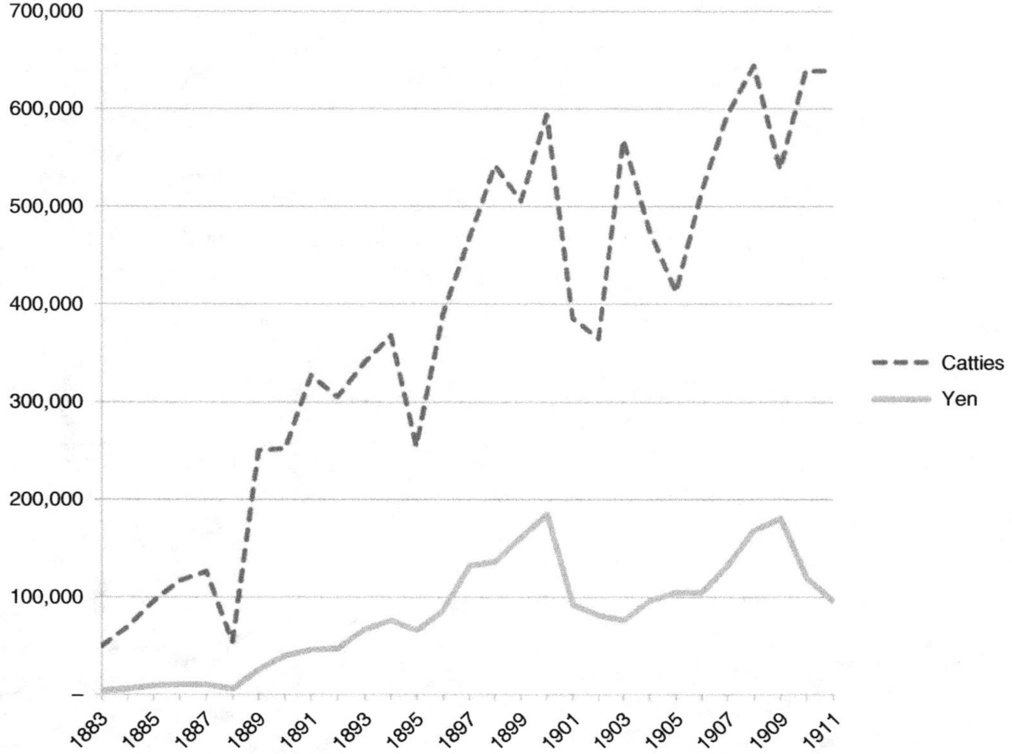

Figure 2.4 Japan plumage exports, 1883–1911
Source: Tōyō Keizai Shimpōsha, *Nihon bōeki yōran* (1 catty = approx. 600 g)

albatrosses covered the surface of the island like snow drifts or fields of cotton[97]. Tamaoki's hunters killed roughly 100,000 albatrosses over the course of the 1888–1889 winter nesting season. The process was then repeated annually. By one estimate, between 1888 and 1902 Tamaoki's settlers killed approximately 5 million birds on Torishima, an island with an area of less than five square kilometres. By the end the slaughter reached such a scale that a railway track was even installed to transport the boxes of wings from the killing fields in the centre of the island to the port.[98]

Tamaoki's success triggered a rush to stake out claims to other deserted islands in the seas around Japan. Tamaoki himself expanded his bird-hunts to other island groups such as the Daitō Islands, and eventually branched out into sugar cultivation and guano mining as well.[99] In 1891 Yorioka Shōzō, another veteran of the *Meiji Maru*'s voyage, opened a sulphur mine on the newly annexed island of Iwo Jima.[100] In 1896 Koga Tatsushirō leased the Senkaku Islands (Figure 2.5), also recently annexed, in order to cull albatrosses and fish for turban snails (see Chapter 7). And in 1905, at the height of the Russo-Japanese War, a sea-lion hunter by the name of Nakai Yōzaburō convinced the government to annex Takeshima (K: Dokdo), midway between Japan and Korea.[101] All of these claimants followed Tamaoki's strategy of promising to establish permanent colonies on bird islands in return for title or leasehold to the islands themselves.

Another of these claimants was a man named Mizutani Shinroku. Like Tamaoki, Mizutani was a Bonins-based sundries merchant with a prior history of *Nan'yō* adventurism. And like Andrew Rosehill, he had spent several years in the 1880s developing trading networks in the Micronesia – at one point being arrested for smuggling by Spanish

97 TMA 617 B4 8: Tamaoki Han'emon, 'Torishima taizai nikki'; Hattori Tōru, 'Bakadori no hanashi', *Dōbutsugaku Zasshi* 12 (15 October 1889): 405.

98 Umō Kenkyūkai, *Umō to negu*, 136–46.

99 Hiraoka, *Ahōdori to 'teikoku' Nihon no kakudai*, 156, 178.

100 Robert D. Eldridge, *Iwo Jima and the Bonin Islands in US–Japan Relations: American Strategy, Japanese Territory, and the Islanders In-Between* (Quantico: Marine Corps University Press, 2014), 38; Ishihara Shun, *Iōtō: kokusaku ni honrō sareta 130-nen* (Tokyo: Chūō Kōron, 2019), 342.

101 NAJ *Dajōkan naikaku kankei kōbun shorui Meiji* vol.29: Meiji 38-nen kōbun shorui No.29; Meiji 38-nen No.1. seikō Teikoku Gikai Gyōsei-ku chihō jichi zassai: 'Okishima o hanaru seihoku 85 ri ni aru mujintō Takeshima to Nageshima Shimane-ken shozoku Okishima shi no shokan to nasu' (28 January 1905).

(a)

(b)

Figure 2.5a and Figure 2.5b Bird-hunting on the Senkaku Islands (c.1939)
Source: Okinawa Prefectural Archives collection

officials in the Carolines. In fact, it was Mizutani's exploits that helped inspire Enomoto and Taguchi to found their own *Nan'yō* trading company in 1890.[102] In 1896, while returning from one of his trading voyages Mizutani was shipwrecked on Marcus. Ever alive to the opportunity this bit of ill-fortune presented, on his return to Japan he set about petitioning the government to annex the island in the hopes of securing a free lease to the island.

Mizutani's petition followed the formula established by Tamaoki. It was true, he admitted, that the immediate business plan relied mainly on hunting albatrosses – an inherently seasonal activity given that the birds only nested on the island in the winter months. But he insisted that this was merely the first stage in an ambitious scheme to establish a community of self-sustaining permanent residents on the island. He had already, he claimed, settled the island with ten men, ten women and three children recruited from the Bonin Islands, who had constructed dwellings for themselves and drilled wells in search of drinking water. If granted a lease, he would set about recruiting more colonists on three-year contracts to work as fishermen on something akin to a share-cropping arrangement: they would be allowed to keep half the market value of whatever they caught. During this time, they would be furnished with food and necessities, and also encouraged to begin 'cultivating' (*kaikon*) Marcus' less than one square kilometre of soil with sugar, tobacco and garden vegetables. Mizutani also promised to station a doctor on the island, provide education for the settlers' children, and also establish a Shintō shrine and, when the time came, a graveyard.[103]

The Marcus Island Incident

On 24 July 1898 the government duly made a formal announcement incorporating Marcus within the Ogasawara municipality of the Tokyo metropolitan government. The timing of the annexation is striking, for Japanese diplomats were well aware that to assert sovereignty over Marcus Island would be a much more delicate matter than had been the case for Torishima. Torishima lay within sea-lanes

[102] Peattie, *Nan'yō*, 16–18.
[103] NAJ, Main building 2A-011-00 No. 00805100: 'Minami-torishima o Tōkyō-fu shozoku Ogasawarajima-shi no shokan to nasu'.

connecting Tokyo to the Bonin Islands, and the risk that a foreign power might object to it being annexed by Japan was slim. Marcus Island was different. It lay more than 1,200 kilometres to the east of the Bonins, and claiming it would extend Japanese sovereignty far into the Pacific Ocean.

Moreover, the annexation was announced in the middle of the Spanish–American War, just as Japan's trans-Pacific neighbour was starting to fully embrace the global culture of imperialism. America's territorial empire in the Pacific predated 1898, of course. The US had possessed a Pacific coastline since the mid-nineteenth century, when the federal government bought Oregon from Britain (1846), seized California from Mexico (1848) and purchased Alaska from Russia (1867).[104] But the war of 1898 generated a new, more explicit way of talking about empire, with politicians and journalists now talking unabashedly of a 'Greater America' that would extend beyond the limits of the continental United States.[105] It also increased the country's Pacific footprint dramatically. Only a month before Japan annexed Marcus Island the US Navy had seized the Spanish colony of Guam (Chamorro: Guåhån). Two weeks before, Congress had voted to annex Hawai'i, at the behest of the same band of renegade American beachcombers who had overthrown the Hawaiian monarchy.[106]

As well as helping stoke a culture of American imperialism, the war also generated a whole new set of strategic imperatives. In the Philippines the US occupiers inherited from Spain a protracted anti-colonial insurgency that would take a decade and great savagery to suppress.[107] And, despite its spectacular conquests, the US Navy was in fact hard-pressed to project maritime military force in the region, at least until the completion of the Panama Canal allowed access to the Pacific. These geopolitical considerations were tinged with a deeply felt racial angst on the part of the Anglo ruling class. Theodore Roosevelt, in particular, was preoccupied by the thought that white settler societies on the Pacific rim were at risk of being overwhelmed by migration

104 Lafeber, *The Clash*, 3, 7.

105 Daniel Immerwahr, *How to Hide an Empire: A History of the Greater United States* (New York: Farrar, Straus and Giroux, 2019), 74–5.

106 Akira Iriye, *Pacific Estrangement: Japanese and American Expansion, 1897–1911* (Cambridge, MA: Harvard University Press, 1972), 51–2.

107 Paul A. Kramer, *The Blood of Government: Race, Empire, the United States, & the Philippines* (Chapel Hill, NC: University of North Carolina Press, 2006).

from the populous Asiatic nations. And Japan was a special object of concern, for it appeared to have growing military as well as demographic clout. In 1898 Roosevelt had argued for Hawaiian annexation by emphasising the strategic threat posed by Japanese migration to the islands. After he helped to broker Japan's peace with Russia in 1905, he would worry privately that it 'might get the "big head" and enter into a general career of insolence and aggression'.[108]

The American reaction to Japan's Marcus annexation betrays some of these anxieties, with the *New York Herald* and other US newspapers reporting that Japan had 'seized' an American island in the Pacific. US naval officials also privately expressed alarm. Shortly after the conclusion of the war with Spain work had begun on a trans-oceanic telegraph that would connect San Francisco, Honolulu and Manila, a project that focused attention on atolls that might serve as cable-landing stations. The Naval Ministry had identified Marcus Island as one such possible site and urged the State Department to protest the Japanese annexation due to 'its previous occupation by an American citizen who still had vested property rights therein'.[109]

It was this culture of imperialism that Rosehill sought to exploit when seeking diplomatic support for his effort to reclaim Marcus for himself.[110] By this point he was working as assistant harbour master of Honolulu, a role that gave him enviable connections to Hawai'i's business community.[111] In early 1901 he banded together with some local investors to deposit a $50,000 bond with the US State Department. He then announced his intention to return to the island to conduct another chemical analysis of its guano deposits on behalf of his newly formed Marcus Island Guano Co.

The expedition was in part a carefully calibrated publicity stunt, aimed at mobilising the US government to back Rosehill's claim. Only after news circulated that a Japanese frigate had been dispatched to intercept the *Whalen*, for instance, did the company's backers announce that there were muskets onboard the ship, ratcheting up

108 Charles E. Neu, *An Uncertain Friendship: Theodore Roosevelt and Japan, 1960–1909* (Cambridge, MA: Harvard University Press, 1967), 18.
109 NARA-II RG 59 Letters Received Relating to Guano Islands, 1852–1905: Acting Secretary of Navy Dept to Secretary of State, 3 August 1899.
110 'Want Cable at Marcus Island', *The Hawaiian Star* (10 July 1902).
111 NARA-II RG 59 Letters Received Relating to Guano Islands, 1852–1905: affidavit by JNO H Soper, JN Fisher, William O Smith, Jr Stackable et al (2 January 1902).

the drama to create the aura of an international showdown. To those who felt that war over a speck of rock in the ocean might be an overreaction, *The Hawaiian Star* pointed out that Japan's annexation threatened to intercept the planned American cable route.[112] Others played on the fact that much of America's Pacific Empire was also sparsely inhabited (according to the 1900 census the entire territory of Alaska had a population of only 68,000, of whom less than a third counted as 'native white').[113] *The Hawaiian Gazette* warned ominously that 'there are hundreds of unoccupied islands in the Aleutian group which have never been occupied by men and [the US] could lose the title to these by Japanese settlement and claim for any Japanese who should choose to occupy them'.[114] In short, failing to uphold Rosehill's claim could result in the erosion of US sovereignty across the Pacific more generally.

Knowing all this, the Japanese ambassador to Washington, DC, could be forgiven for feeling a degree of trepidation. It was true that Japan had successfully negotiated an end to extraterritoriality in 1894: in that sense, the clash over Marcus was fundamentally different to earlier disputes over the rights of American beachcombers on the Bonin Islands.[115] Still, the US government had a track record of bellicosity when it came to asserting claims to guano islands. In 1859 it had dispatched a frigate to support an American citizen's claim to the island of Navassa, off the coast of Haiti.[116] And it was, after all, a false rumour about an attack on an US battleship in Havana harbour that had sparked the Spanish-American War.

But the ambassador managed to convince the State Department to issue a letter to Rosehill, ordering him not to resort to force of arms and instead let the question of Marcus' sovereignty be resolved through diplomatic channels. The letter was cabled to Tokyo and handed to the captain of the fast cruiser *Kasagi*, which promptly set off to intercept the *Whalen*. The *Kasagi* did beat the Whalen to Marcus, but it turned out there was no

112 'Want Cable at Marcus Island'.

113 US Bureau of the Census, *Twelfth Population Census of the United States – 1900* (Washington DC: US Government Printing Office, 1901), 482.

114 Kokusai Kōhō Kenkyūshitsu, *Minami-torishima jiken: Hōka Daigaku kokusai kōhō enshū hōkoku* (Tokyo: Tōkyō Teikoku Daigaku Kōnai Kokusai Kōhō Kenkyūshitsu, 1902), 10.

115 Louis G. Perez, *Japan Comes of Age: Mutsu Munemitsu and the Revision of the Unequal Treaties* (Vancouver: Fairleigh Dickinson University Press, 1999).

116 Kokusai Kōhō Kenkyūshitsu, *Minami-torishima jiken*, 12.

harbour large enough to accommodate it. It waited off-anchor, burning coal to remain in position until, running low on fuel, it was forced to return to Yokosuka.[117] So when Rosehill finally arrived he was met not by a Japanese warship but by a more modest detachment of marines camped out on the island.

This set the stage for an uneasy standoff. The Japanese commanding officer, a Lieutenant Akimoto, handed over the consular note and forbade Rosehill or his crew to set foot on the island. But mindful that the final resolution of the sovereignty dispute was as yet uncertain, he did permit the two scientific members of the expedition to make landfall to conduct a survey of the island.[118] The men spent a week traipsing around the island, assessing the value of its phosphate reserves and observing the atoll's geology and ecology. At this point Rosehill informed Akimoto that his team required more time to complete its survey. As he later recounted:

> The lieutenant shrugged his shoulders, glanced significantly at the sixteen armed marines drawn up in line and indicated that he would be compelled to enforce the orders left with him . . . not to permit the Captain and his party to land or to remain on the island. As there were sixteen men beside the lieutenant and about fifty stalwart Japanese fishermen, [I] concluded that [my] instructions from the company which were to yield to a superior force, required [me] to depart.[119]

A disconsolate Rosehill weighed anchor and returned to Honolulu. On his return he tried to convince the State Department to defend his claim to the island, by dispatching a warship if necessary. But the department's officials showed little interest. Indeed, we might possibly infer slight sense of relief that the *Kasagi* had arrived first, managing to nip the dispute in the bud. With no diplomatic assistance forthcoming, Rosehill petitioned the State Department to at least reimburse the Marcus Island Guano Co. for the expenses it had incurred on the *Whalen*'s fruitless voyage.[120] Eventually he received a curt notice pointing out that although he, Rosehill, had indeed registered his claim to Marcus back in 1889, he had neglected to pay the purchase

117 *Ibid*, 7.
118 'Minami-torishima jiken no kaiketsu' *Asahi Shimbun* (6 September 1902).
119 Kokusai Kōhō Kenkyūshitsu, *Minami-torishima jiken*, 8–9.
120 'Uncle Sam Will Not Fight Over Marcus Island', *The Hawaiian Star* (21 July 1902).

a bond until January 1902 – four years after the Japanese annexation of the island and only six months before he set off in the *Whalen* to reclaim it.[121]

For the State Department this supplied a neat post hoc justification for inaction. But to suggest that the outcome of the incident was determined by the neutral application of constitutional principles would be misleading. As we have seen, the US government was notoriously inconsistent in its interpretation of the Guano Islands Acts. Only a few months earlier, after all, it had happily issued Rosehill a certificate vesting in him 'the usual rights and privileges as … discoverer' of Marcus Island.[122] Only after the confrontation with Japan did it reject his claim. Indeed, the urgency with which both American and Japanese diplomats responded to Rosehill's expedition reveals the scope for contingency in the whole affair. If Rosehill's backers had secured more public support for their campaign to retake the island, things might have played out very differently indeed. And if he had beaten the *Kasagi* to Marcus and confronted Mizutani's bird-hunters directly, the situation might have escalated very quickly.

On a deeper level, too, it was Mizutani's performance of occupation that carried the day. He had managed to establish a colony on the island, whereas Rosehill's first attempt at colonisation had foundered and his second one come too late. In this sense it is worth noting the degree to which the timing of the dispute played into Mizutani's hands, for his colony on Marcus Island turned out to be a fragile thing indeed. By 1902 his hunters had already decimated the island's albatross population and were now reduced to culling sooty and white terns, smaller and therefore less valuable birds which only nested on the island in the summer months.[123] In later years, as the bird population dwindled further, Mizutani diversified into guano mining, and did manage to extract a moderate amount of phosphate through to the end of World War I (peaking at slightly over 10,000 tons in 1913). But the mine was

121 NARA-II RG 59 Letters Received Relating to Guano Islands, 1852–1905: 'Marcus or Weeks'.

122 NARA-II RG 59 Letters Received Relating to Guano Islands, 1852–1905: 'In the Matter of the Application of Andrew A. Rosehill for the Right to Remove Guano Deposits from Marcus Island' (20 January 1902).

123 William Alanson Bryan, 'A Monograph of Marcus Island: An Account of Its Physical Features and Geology, With Descriptions of the Fauna and Flora', *Occasional Papers of the Bernice Pauahi Bishop Museum* II no. 1 (1904): 97–8, 104.

shuttered during the recession of 1920, never to reopen.[124] Marcus may remain Japanese sovereign territory to this day, but Mizutani ultimately failed to deliver on his promise to establish a permanent colony on the island.[125]

None of this was evident to US diplomats in the summer 1902, nor to the crew of the *Whalen* who had sailed halfway across the world's largest ocean only to return in frustration. Instead, they saw a young, vigorous Asiatic empire encroaching on the United States' oceanic territory, colonising where America had failed to colonise. And they saw environmental destruction on an appalling scale: the wholesale massacre of innocent birds for vulgar commercial gain. The solution to both these problems lay in nature conservation.

124 Asō Yawata, *Rinkō* (Tokyo: Maruzen, 1940), 73–4.

125 Like the Bonin Islands, Marcus Island was occupied by the US military after World War II, but was returned to Japanese sovereignty in 1968. For further discussion of the post-war US occupation of Japanese island groups, see Chapter 5.

3 *Bird and Sovereignty Conservation in the Northwest Hawaiian Islands, 1898–1911*

The story of the Marcus Island colony of goonies is one of death and extermination .... The long wing feathers of all the birds were pulled out [by the Japanese hunters] and carefully preserved to be shipped to America and Europe and sold as 'eagle feathers', which were in great demand for trimming on ladies' hats. The feathers from the breast were plucked off and sold by the pound. A profitable business was thus developed, with the deplorable result that within six years the entire colony of these splendid birds has been exterminated.[1]

So wrote William Allison Bryan, a member of the ill-fated American expedition to Marcus Island in the summer of 1902. An ornithologist affiliated with Honolulu's Bishop Museum of Natural History, Bryan was one of only two members of the expedition that the Japanese Navy had allowed to set foot on the island so as to make scientific observations. Together with a chemist tasked with assessing the value of Marcus' guano reserves, Bryan spent a week traipsing around the island. As a man of science, he clearly cut an impressive figure. In the words of one Japanese bird-hunter, he was 'quite unlike that vulgar swindler Rosehill, he had a gentlemanly demeanour ... he rushed about the island day and night conducting his investigations with the utmost diligence; we were all surprised at how much he exerted himself in his research.'[2]

As well as regarding the Japanese slaughter of Marcus' 'goonies' (albatrosses) with unvarnished horror, Bryan was disturbed by the thought that Marcus Island, once American territory, had been lost to the Empire of Japan. In the decade to come he would become a key figure in a campaign to repel Japanese bird-hunters from the many uninhabited islands dotted across the North Pacific that were, as he understood it, sovereign territory of the United States. To this end he mobilised a broad

[1] Bryan, 'Monograph of Marcus Island', 106.
[2] 'Minami-torishima jiken no kaiketsu', *Asahi Shimbun* (6 September 1902).

network of allies, from naval officers to conservationists to the President of the United States himself, motivated to varying degrees by concern for the fate of birds and a determination to conserve American sovereignty in the Pacific. To do so they pioneered new techniques for excluding alien nationals from putatively American territory, and transformed a scattering of abandoned guano islands into what would eventually become the Papahānaumokuākea Marine National Monument, one of the largest national parks in the world.

Race, Nature and Sovereignty in the Post-1898 Pacific

For a glimpse into the worldview that animated Progressive-era America, the mind of Theodore Roosevelt makes a good place to start. In one sense, no individual did more to shape American politics and society from the 1890s through to World War I, whether in the realms of domestic reform or foreign policy. In another, Roosevelt's successes were achieved precisely because his beliefs were so widely shared by a broad cross-section of American elites. And though a bookshelf of biographies attests to Roosevelt's fascinating and complex character, he cared for two things above all else: empire and nature.[3]

No one was a more vigorous cheerleader for American imperial expansion than Roosevelt. In 1898 he had quite literally led the charge into battle, storming Spanish defences in Cuba at the head of a volunteer militia unit.[4] And as president he made strenuous efforts to build up the fleet size of the US Navy, keenly aware that America's new Pacific Empire was acutely overextended.[5] Roosevelt's understanding of geopolitics was also inflected by Social Darwinist anxiety. He was especially preoccupied by the thought that white settler societies risked being overwhelmed by migration from the populous Asiatic nations, and by Japan in particular. In 1898 had Roosevelt argued for Hawaiian annexation by emphasising the strategic threat posed by Japanese migration to the islands. After he helped to broker the end

3 For more recent works that focus on Roosevelt as conservationist, see Ian Tyrell, *Crisis and the Wasteful Nation: Empire and Conservation in Theodore Roosevelt's America* (Chicago: University of Chicago Press, 2015); Michael R. Canfield, *Theodore Roosevelt in the Field* (Chicago: University of Chicago Press, 2015).

4 William O. Walker, *National Security and Core Values in American History* (Cambridge: Cambridge University Press, 2009),

5 Neu, *Uncertain Friendship*, 90.

of the Russo-Japanese War in 1905, he worried privately that Japan 'might get the "big head" and enter into a general career of insolence and aggression'.[6]

Moreover, Japanese immigration to the continental US, much of it routed through Hawai'i, led to a nativist backlash that in turn morphed into a major diplomatic crisis that consumed the last years of Roosevelt's presidency. When the San Francisco Board of Education attempted to segregate the city's schools, Roosevelt found himself caught between anti-immigrant sentiment on the West Coast and a Japanese government determined that its citizens not be targeted for exclusion in the same way that Chinese immigrants had been in 1881. The dispute was eventually defused by the 1907 Gentlemen's Agreement, whereby Japan took upon itself the task of restricting emigration to the mainland US rather than submit to the indignity of an explicit immigration ban.[7] But at one point Roosevelt felt relations to be so tense that he despatched the US fleet on a round-the-world voyage via the Pacific, hoping to display American military might for the benefit of Japan audiences.[8]

Whist pressing a strenuous pro-imperial foreign policy abroad, Roosevelt was also a keen advocate of nature conservation at home. As president from 1901 to 1909, Roosevelt inaugurated the federal government as a force for nature conservation. Some of his innovations were concerned with sustainable resource use in a straightforwardly instrumentalist sense, such as appointing Gifford Pinchot as the first chief of the Forest Service and organising the first ever Governors' Conference on the Conservation of Natural Resources.[9] But Roosevelt also conceived of nature not only as a material resource, such as timber, but also as an aesthetic, indeed spiritual one. During his presidency he made sweeping use of his executive powers to establish a broad array of national forests and wildlife reserves within the metropolitan United States and its colonies.[10]

A number of things are worthy of comment here. The first is that Roosevelt did not conceive of these parks in isolation; rather, he did so

6 *Ibid*, 18. 7 *Ibid*, 43–4. 8 *Ibid*, 116–17.

9 Karl Jacoby, *Crimes Against Nature: Squatters, Poachers, Thieves, and the Hidden History of American Conservation* (Oakland: University of California Press, 2001), 202.

10 Roderick Nash, *Wilderness and the American Mind* (New Haven: Yale University Press, 1982); Donald Worster, *Nature's Economy: A History of Ecological Ideas* (New York: Cambridge University Press, 1994), 263.

in response to campaigns by a burgeoning network of American conservation activists. One key ally was the zoologist William Hornaday, who founded the Bronx Zoo and led a crusade to protect America's few remaining bison herds.[11] Bird conservationists constituted another key constituency. During the 1890s a network of Audubon societies had sprung up across the nation, dedicated to stemming the devastation inflicted on wild bird populations by the global plumage trade (see Chapter 2). Audubon societies campaigned on a number of fronts. At a national level they lobbied for a ban on feather exports and other legislation to restrict the plumage trade, and to raise public awareness for the plight of wild birds. As a result of Audubon campaigning, for instance, several states introduced 'bird days' into their school curricula.[12] Local Audubon branches also campaigned for particular breeding or migration sites to be placed off-limits to commercial hunters. Of the eighty-eight nature reserves that Roosevelt established, thirty-three were designated specifically for bird protection.[13]

The second point is that these wildlife conservation campaigns, while expressed in terms of genuine concern for the birds themselves, were generally more about conserving *wilderness* than conserving *life*. Unlike contemporary movements such as the Theosophists, for example, Audubon clubs did not express any interest in vegetarianism.[14] Concern over wilderness was in turn bound up with other preoccupations. One was gender. Roosevelt and many other conservationists understood the wilderness to be a place to hone virility through hiking, camping and amateur hunting, which (unlike commercial hunting) was deemed to be a legitimate pursuit. Indeed, a paradox of bird conservation in this period was that the ornithologists who promoted it also understood the killing and collecting of bird carcasses as integral to their own scientific endeavours.[15] The wearing of plumage, on the other hand, was not only

11 Jacoby, *Crimes Against Nature,* 96.
12 Doughty, *Feather Fashions and Bird Preservation*, 31–2.
13 Based on executive orders compiled by the UC Santa Barbara 'American Presidency Project': www.presidency.ucsb.edu/documents/app-categories/written-presidential-orders/presidential/executive-orders. Last accessed 9 February 2022.
14 Helena P Blavatsky, *The Key to Theosophy*: (Global Grey, 2013), 157; H. Augusta Howard, 'The Audobon [sic] Society', *Food, Home and Garden* (March 1897): 68–9.
15 Mark Barrow, *A Passion for Birds: American Ornithology After Audubon* (Princeton: Princeton University Press, 2000).

cruel but indicated an inappropriately frivolous type of femininity: complaints about the 'wanton' harvesting of feathers also implicitly questioned the character of the sort of woman who wore them.[16]

Wild bird protection was also a largely upper-class movement. Audubon members generally hailed from the wealthier ranks of society, as hinted at by members' pledges not only to 'refrain from wearing the feathers of any birds not killed for purposes of food' – but also to discourage their friends *and also servants* [my italics] from doing likewise.[17] And the new national parks generally catered to the sort wealthy tourists who could afford to take prolonged vacations in relatively remote locales, and who viewed physical exertion as leisure not drudgery. But policing the parks in practice meant designating them off-limits to hunters from local communities.[18]

Finally, concern for wilderness was also tied up to anxiety about race and nation. In this sense, American wildlife conservationists were participating in a much broader trans-national conversation. It was around this time that German nationalists began tracing the roots of the Teutonic *volk* back to the primeval Hercynian Forest, from which Arminius had led the barbarian hordes to do battle against the Romans (i.e. the French).[19] In a similar vein, the novelist Walter Scott located the source of Scottish culture in the rugged wilderness of the highlands.[20] As descendants of a settler colonial society, Anglo-Americans could hardly claim a national identity rooted in indigeneity. So instead they developed an idiosyncratic inversion on the theme whereby national character was forged through dialectical engagement with supposedly uninhabited space.[21] The most famous expression of

16 Doughty, *Feather Fashions and Bird Preservation*, 63; Jennifer Price, *Flight Maps: Adventures in Nature in Modern America* (New York: Basic Books, 1999), 81–2.

17 Doughty, *Feather Fashions and Bird Preservation*, 31, 96–7.

18 Jacoby, *Crimes Against Nature*, 33, 40–1.

19 Simon Schama, *Landscape and Memory* (New York: Vintage, 1996), 100–2; Michael Imort, 'A Sylvan People: Wilhelmine Forestry and the Forest as a Symbol of Germanism' in Thomas Lekan and Thomas Zeller, *Germany's Nature: Cultural Landscapes and Environmental History* (New Brunswick: Rutgers University Press, 2005): 55–80.

20 Hugh Trevor-Roper 'The Invention of Tradition: The Highland Tradition of Scotland', in *The Invention of Tradition* ed. Eric Hobsbawm and Terence Ranger (Cambridge: Cambridge University Press, 2012).

21 Frederick Jackson Turner, 'The Significance of the Frontier in American History'; William Cronon, 'Revisiting the Vanishing Frontier: The Legacy of

this was Turner's Frontier Thesis, which argued that the dialectic between (white) civilisation and wilderness was the vital source of the American democratic spirit.

As William Cronon has shown, wilderness conservation in practice often meant cordoning off space for enjoyment by wealthy white men at the expense of people of colour.[22] In perhaps the bitterest irony, Native Americans were sometimes evicted from newly created national parkland for failing to adhere to the requisite tropes of noble savagery.[23] But immigrant groups were also frequent targets of nature conservation campaigns: across the continental United States bird conservationists waged campaigns to criminalise Hispanics and Italian immigrants for engaging in their customary hunting behaviour.[24] In sum, the cult of wilderness legitimated expelling people from land in the name of nature conservation.

The new US colony of Hawai'i provided an obvious space for Rooseveltian preoccupations with nature and empire to mingle. After 1898 a wave of Americans descended on the archipelago, and along with the administrators and naval officers came also a host of taxonomical carpetbaggers, ethnographers and naturalists who sought both to catalogue and to steward the natural riches of the new US territory. The American colonisers arrived in Hawai'i to encounter a multi-racial society unlike anything they would have known back home.

Most familiar was Hawai'i's substantial white population, who mainly descended from missionaries, sugar planters and beachcombers such as Andrew Rosehill. Even before annexation whites had established themselves as a significant demographic (and political) presence within Hawai'i, comprising about 20 per cent of the total population according the census of 1900.[25] That same census classified another

Frederick Jackson Turner', *The Western History Association* 18 no. 2 (April 1987).

22 Cronon, 'The Trouble with Wilderness'; Donna Haraway, 'Teddy Bear Patriarchy: Taxidermy in the Garden of Eden, New York City, 1908–1936', *Social Text* 11 (1984); Gray Brechin, 'Conserving the Race: Natural Aristocracies, Eugenics, and the US Conservation Movement', *Antipode* 28 (1996).

23 Jacoby, *Crimes Against Nature*, 149–50.

24 Louis S. Warren, *The Hunter's Game: Poachers and Conservationists in Twentieth-Century America* (New Haven: Yale University Press, 1994)

25 US Bureau of the Census, *Twelfth Population Census of the United States – 1900* (Washington, DC: US Government Printing Office, 1901), 498.

20 per cent of the population as Native Hawaiians. Though many Hawaiians either resisted or contested the terms of annexation, the colonisers generally regarded the archipelago's Indigenous population with a mix of curiosity and condescension, assuming them destined for either extinction due to introduced pathogens or assimilation 'when brought into contact with an advanced social order'.[26] One of Bryan's early tasks at the Bishop Museum, for instance, was to compose a diorama from a series of casts taken of 'natives...to illustrate the most important or characteristic employments of old Hawaiians.'[27]

Far more threatening were the Japanese plantation labourers who had since the 1880s been recruited en masse to work in Hawai'i's booming sugar industry, and who by 1900 constituted a plurality (40 per cent) of the total population. And it was not just the demographic strength of the Japanese that was alarming. US annexation had profoundly destabilised the political economy of labour in the Hawaiian archipelago. Until 1898, the Hawaiian legal code's support for indentured labour (shirkers could be and were punished in the courts) had kept Japanese immigrants confined to the plantations.[28] But the 1900 Organic Act that established the Territory of the Hawai'i had abolished indenture, allowing Japanese workers to leave their plantations and take up alternative work elsewhere. Japanese people soon began making inroads into areas of the economy formerly dominated by whites and Native Hawaiians, such as construction, stevedoring and fishing.[29]

[26] Katharine Coman, *The History of Contract Labor in the Hawaiian Islands* (New York: Macmillan, 1903), 1–2. For the myth of the fatal impact see Kerry R. Howe, *Nature, Culture, and History: The 'Knowing' of Oceania* (Honolulu: University of Hawai'i Press, 2000), 43–4. On the demographic, political and cultural impact of introduced diseases, see Seth Archer, *Sharks Upon the Land: Colonialism, Indigenous Health, and Culture in Hawai'i, 1778–1855* (Cambridge: Cambridge University Press, 2018). For the history of Native Hawaiian resistance to annexation, see Noenoe K. Silva, *Aloha Betrayed: Native Hawaiian Resistance to American Colonialism* (Durham, NC: Duke University Press, 2004).

[27] William T. Brigham, *Director's Report for 1902* (Honolulu: Bishop Museum Press, 1903), 6–7.

[28] Ronald T. Takaki, *Pau Hana: Plantation Life and Labor in Hawaii, 1835–1920* (Honolulu: University of Hawai'i Press, 1983), 148.

[29] Alan Takeo Moriyama, *Imingaisha: Japanese Emigration Companies and Hawaii, 1894-1908* (Honolulu: University of Hawai'i Press, 1985), 137; Manako Ogawa, *Sea of Opportunity: The Japanese Pioneers of the Fishing Industry of Hawaii* (Honolulu: University of Hawai'i Press), 37–8.

By granting workers the right to unionise the Organic Act also had the unintended consequence of unleashing a wave of strike activity on sugar plantations across Hawai'i. On the day the act passed, Japanese labourers marched through downtown Honolulu, chanting 'We are free people!'[30] Throughout the summer of 1900, Japanese cane cutters staged strikes and work stoppages at plantations across Hawai'i, calling for shorter working hours, increased wages and more Japanese promoted to positions as overseers.[31] A decade of industrial unrest ensued, culminating in a general strike of Japanese cane-cutters that managed to shut down every sugar plantation on O'ahu.[32]

All this helped give rise to a pervasive Anglo unease that America's new colony was in danger of being overrun by Oriental hordes.[33] Journalists seized on the fact that some strikers were using rallying calls such as 'Yamato spirit' as evidence of a racial conspiracy. One Honolulu-based solicitor wrote to the State Department warning that 'things are not very steady here owing to the labour situation, Japanese trouble and other minor things ... Since the [Russo-Japanese] war they have gotten the swell head and are literally trying to run things out here.'[34] Some theories verged on the conspiratorial. The head of the US Department of Agriculture's Hawaiian research station wrote to Roosevelt claiming that 'the unrest among these people in Hawai'i was caused and is being fomented and encouraged by official representatives of the Japanese Government who have been sent to Hawai'i for that purpose ...'[35]

The Hawaiian authorities reacted to the newly assertive Japanese presence in Hawai'i in a number of ways. The Honolulu municipality forbade contractors from employing 'Asiatic labour' on road repair crews, and in 1902 the territorial government announced that aliens must pay duty of one cent per pound on fish caught in sea waters off Hawai'i, along with a raft of other laws that targeted Japanese

30 Takaki, *Pau Hana*, 148–9.
31 Coman, *History of Contract Labor in the Hawaiian Islands*, 47–8.
32 Takaki, *Pau Hana*, 150–1.
33 'Trouble Threatened on Maui', *The Hawaiian Star* (12 August 1905).
34 NARA-II MD, State Dept 1906–10, Japanese Immigration Crisis: H. R. Castle to W. S. Rossiter, 5 December 1906.
35 NARA-II MD, Territories Office, Hawai'i, Labor, Jared G. Smith to Wilson, 29 January 1909.

fishermen at the local level.[36] The territorial government also made a game attempt to transform Hawai'i from a society of large plantations reliant on Asian labour to one populated by (white) Jeffersonian homesteaders. Hoping 'to offset the large percentage of Japanese labourers now within the Territory by a desirable class of Europeans eligible for citizenship', the Board of Immigration despatched agents to scour the more impoverished corners of the Old World, enticing emigrants with subsidised passage and land grants on arrival.[37] For their part, sugar planters fought back against the militancy of the cane-cutters' unions by recruiting workers from Korea and the Philippines, trying to undermine worker solidarity through racial diversification.[38]

William Alanson Bryan (Figure 3.1a) found himself square in the middle of all of this. A devout Quaker sired in the nation's heartland, Bryan had trained in zoology at Iowa State College before arriving in Honolulu in 1900 to take up an appointment as resident taxidermist at the Bishop Museum of Natural History, as well as 'Special Inspector, Birds and Animals' – deputised by Roosevelt himself no less.[39]. Immediately upon arrival, he threw himself into the public life of the United States' new colony. At the Bishop Museum, he used cutting-edge taxidermic methods, 'the same as those used with such capital success in South Kensington, New York and other great museums', to stage elaborate dioramas of native Hawaiian birds mounted against 'their natural surroundings'.[40] He also worked with the Hawaiian Sugar Planters' Association to introduce pest-eating bird species to plantations, and established a Pacific Science Institution under whose auspices he would hopscotch the Pacific, travelling as far as Easter

36 Ogawa, *Sea of Opportunity*, 56–7.

37 HSA Governor Frear, Territorial – Immigration 1907–1909: 'Immigration'.

38 Takaki, *Pau Hana*, 161–3. As Takaki describes, this strategy was only temporarily effective. Within just over a decade Japanese and Filipino labour organisers succeeded in coordinating an interethnic strike among cane-cutters.

39 NARA-II MD, Territories Office, Hawai'i, Bird Reservations: Interior Secretary to Assistant Secretary of the Navy, 1 December 1904. The Bishop Museum had been established by the banker Charles Bishop in 1889 in memory of his late wife Princess Bernice Pauahi Bishop, the last legal heir of the Kamehameha Dynasty. Its mandate was 'to preserve and exhibit to all who care to look relics of her people and the kindred races of the Pacific Ocean'.

40 William T. Brigham, *Director's Report for 1902* (Honolulu: Bishop Museum Press, 1903), 7.

Figure 3.1a William Alanson Bryan and his wife at home in Honolulu (c.1903)
Figure 3.1b Bryan's wife with their Japanese maidservant (c.1903)
Figure 3.1c Max Schlemmer's son carrying a brace of birds (c.1903)
Source: University of Hawai'i at Mānoa William Alanson Bryan collection

Island to investigate its potential as a sheep ranch.[41] In 1914 he even ran for the Hawaiian governorship.[42]

Bryan was starkly aware of the Japanese presence in Hawai'i. At home he employed a Japanese maidservant in his own household (Figure 3.1b), and at work he would have encountered Japanese on a daily basis.[43] In 1902 the director of the Bishop Museum noted with alarm the uptick in Japanese plantation workers visiting the museum.[44] Concerned that 'the lawless element of the community is increasing', the museum began placing exhibits in glass cases for protection, including Bryan's elaborate dioramas of stuffed birds.[45] Bryan himself was anxious enough about the Japanese to write a lecture on the 'world crisis' engendered by 'the collision of races ... evenly matched' in the Pacific. Integration offered no solution, for as a keen eugenicist Bryan believed that miscegenation between whites and Japanese would result in offspring 'apt to be physically inferior to the average of either parent stock'. Indeed, 'the preventing of contact of this character is only practicable when measures are taken to prevent the people of one race entering the territory of the other for economic reasons'.[46]

Bryan's attitude to the Japanese presence in Hawai'i was particularly shaped by his voyage to Marcus Island in the summer of 1902. He would later describe the landing on Marcus as 'the most exciting experience of my life', and recounted his disappointment that 'when our little company of ocean-tossed explorers reached the island we had always called our own, it was to find the Japanese flag boldly flying over the tree tops from a pole a hundred feet high ... and a band of half-dressed coolies scurried about like crabs on the beach.

41 UHM WA Bryan Collection, Bryan misc correspondence: Brigham to Bryan (25 January 1909); Bryan misc correspondence: HSPA Assistant Secretary to Bryan (9 December 1914); NARA-II MD, Territories Office, Hawai'i, Education: 'Pacific Science Institution'.

42 UHM William Alanson Bryan Collection, Bryan scrapbook 1912-: 'Professor Bryan Home: Confident of Chances for Governorship'; 'Pinkham Next Governor of Hawai'i: Nomination Made by President Wilson Receives the Approval of the Senate'.

43 UHM WA Bryan Collection, Bryan photo album.

44 William T. Brigham, *Director's Report for 1901* (Honolulu: Bishop Museum Press, 1902), 9–10.

45 Brigham, *Director's Report for 1902*, 10.

46 UHM WA Bryan Collection, 'Race Contact in the Pacific' [m.s.] (19 May 1911).

It was obvious that we were both anticipated and expected.'[47] As extracts like these suggest, his whole account of the expedition was suffused with the frisson of colonial rivalry.

Despite the failure of Rosehill's initial expedition, Bryan was initially confident that the island would eventually be restored to American sovereignty. After returning to Honolulu, he lent his support to Rosehill's campaign to have the US government intercede with Tokyo on his behalf, for instance petitioning Governor Dole to raise the issue with Washington.[48] But as the suit ground on, Bryan's confidence in a favourable outcome waned. In late 1903, when the State Department rejected Rosehill's claim decisively, Marcus Island now morphed into a symbol of crumbling American sovereignty in the Pacific. In an essay on 'America's Lost Possessions' Bryan wrote that 'what happened at Marcus Island is part of the same story told of [guano islands such as] Palmyra and Jarvis and Baker. All of them were American possessions once.'[49]

Protecting the Northwest Hawaiian Islands

In this new, more pessimistic mood, Bryan's attention shifted to the United States' other deserted islands in the Pacific – islands that were also in danger of being lost to Japanese encroachment, he fretted. Of particular concern were the outlying Northwest Hawaiian Islands, a chain of atolls and volcanic outcrops stretched over a thousand miles of ocean.

Many of these islands had once been well-known to Native Hawaiians, for they 'were considered 'äina akua, or ancestral islands where the souls of the deceased would travel to and lived in afterlife'.[50] Prior to the collapse of the kapu system in 1819, the easternmost islands in the group were visited regularly.[51] Mokumanamana

47 BPBM WA Bryan collection: William Alanson Bryan, 'America's Lost Possessions' [m.s.] 82.
48 BPBM WA Bryan collection: Dole to Bryan, 19 September [n.d.]; 'Want Cable at Marcus Island'.
49 BPBM WA Bryan collection: William Alanson Bryan, 'America's Lost Possessions' [m.s.].
50 Kikiloi, 'Rebirth of an Archipelago', 89.
51 Evidence of Native Hawaiian activity on the westerly islands is circumstantial at best. But skeletons of flightless Laysan ducks have been recovered on Lisianski Island, and Mark Rauzon has speculated that the ducks may have been introduced to Lisianski by Hawaiian voyagers. See Office of Hawaiian Affairs,

(Necker Island), as the 'axis point between the world of the supernatural and the living', was used to stage calendrical rituals that conferred mana on Hawaiian ali'i.[52] Nihoa shows signs of continuous habitation during the same period, and was most likely used as a staging post to access Mokumanamana. Until as late as the 1840s Native Hawaiians were using feathers harvested on Nihoa to decorate kahili (royal standards) and other ritual artefacts.[53] In 1857 Kamehameha IV had (re) annexed the entire island chain for the Kingdom of Hawai'i – this time not to harness their ritual power or steward the souls of the dead but to cash in on the global guano rush.[54] In the event, though, few attempts were made to mine the islands, and by the time the US annexed Hawai'i in 1898 they were almost entirely uninhabited.

The first hint that Japanese bird-hunters were operating on the Northwest Hawaiian Islands dates to 1900, when the Japanese consul at Honolulu wrote to Governor Sanford Dole enquiring whether his government might lease them from the Territory of Hawaii so as 'to secure permission or right of fishing and bird-catching by the Japanese subjects'. Dole declined the request, citing both 'sentimental' and

Native Hawaiian Claims to the Lands and Resources of the Northwest Hawaiian Islands (February 2003), 18.

52 Kekeuwa Scott T. Kikiloi, Kūkulu Manamana: Ritual Power and Religious Expansion in Hawai'i – the Ethno-Historical and Archaeological Study of Mokumanamana and Nihoa Islands' (PhD diss. University of Hawai'i at Mānoa, 2012), 146–65, 316; Paul L. Cleghorn, 'The Settlement and Abandonment of Two Hawaiian Outposts: Nihoa and Necker Islands', *Occasional Papers of the Bernice Pauahi Bishop Museum* (1988), 48.

53 David Malo, *Hawaiian Antiquities (Moolelo Hawaii)*, trans. N. B. Emerson (Honolulu: Hawaiian Gazette, 1903), 65.

54 It hard to say definitively when or even if Native Hawaiians ceased to attach ritual significance to the Northwest Hawaiian Islands and the birds that nested on them. But from as early 1825 members of Hawaiian royalty were buried in the Christian style in self-consciously Western mausoleums. This includes Prince Kalaniana'ole (1871–1922), the most prominent champion of Native Hawaiian culture in the immediate aftermath of US annexation, whose personal papers contain no reference to the islands. Well into the twentieth century royal Hawaiian funerals continued to involve copious amounts of featherwork, including the prominent display of kahili. But the ritual association between divine/chiefly power and ocean-going migratory birds such as 'iwa (frigate birds) and ka'upu (black-footed albatrosses) seems to have waned over time. See Ralph Thomas Kam, *Death Rites and Hawaiian Royalty: Funerary Practices in the Kamehameha and Kalākaua Dynasties, 1819–1953* (Jefferson: McFarland & Co., 2017), 33–5; 'The Funeral of Prince Jonah Kūhiō Kalaniana'ole', *Paradise of the Pacific* (February 1922); Stacy L. Kamehiro, 'Featherwork in the Hawaiian Monarchy Period', in *Royal Hawaiian Featherwork*, 83–6.

'commercial' reasons why the birds should be protected. In fact, Dole was concerned less about the hunting of the birds *per se* than about by whom they were to be hunted. As he argued to the US Secretary of the Interior, 'If this business of collecting bird skins is allowed should it not be limited to American citizens or at any rate to the residents of the Hawaiian Islands?'[55]

It so happened that around this time the Northwest Hawaiian Islands, along with the islands of the North Pacific more generally, began attracting interest from other quarters as well. Hawaiian annexation had finally cleared the way for the attainment of one of Washington's long-held strategic goals: the laying of a trans-Pacific telegraph cable. The prospect of a cable in turn dramatically increased the strategic value of remote islands in the North Pacific (something that Bryan made sure to emphasise during his public campaign in support of Rosehill), and in 1900 Secretary of State Hay commissioned a survey of US insular possessions in the Pacific (most of them guano islands that had either been abandoned or never even excavated) to see which could be used as landing stations.[56]

This brought the activities of Japanese bird-hunters into new focus. In 1901, the year before Rosehill's expedition, Mizutani's bird-hunters on Marcus got into an altercation with a visiting US troop transport, even attempting (unsuccessfully) to order its commanding officer off the island.[57] Another dispute broke out over Wake Atoll (Marshallese: Ānen Kio), roughly equidistant between Guam and Tokyo. Wake had been claimed as US territory in 1899, shortly after the invasion of the Philippines, with the explicit aim of using the island as a cable-landing station. In the summer of 1902 captains from two different ships reported that a Japanese settlement had been established on the island, ostensibly for the purposes of 'gathering guano and drying fish'.[58] This prompted a diplomatic exchange between Washington and Tokyo in which the Japanese foreign minister renounced territorial sovereignty over Wake

55 NARA-II MD, Territories Office, Hawai'i, Dole to Interior Secretary (6 July 1903).

56 NARA-II State Department Guano Islands: J. D. Hague to Secretary of State John Hay (9 February 1900).

57 'Jap Squatters May Show Fight: Marcus Island Believed to be in Possession of Little Brown Men', *The Hawaiian Star* (3 July 1902).

58 'Japs Have Island', *The Hawaiian Star* (26 August 1902).

but asked that Japanese subjects on the island 'be properly protected as long as they are engaged in peaceful occupations'.[59]

While this statement certainly settled the question of which government exercised territorial sovereignty over Wake, it was at the same time a fairly unambiguous assertion of Japanese citizens' right to inhabit the islands as long as they were engaged in 'peaceable occupations'. This was perfectly in line with Tokyo's insistence, in the era before the Gentlemen's Agreement, that Japanese nationals be permitted to migrate freely to Hawai'i and to the continental United States. The State Department did not object, and in any event Wake was not selected as a cable station, which put an end to the matter. Only months later a US naval captain observed that another band of Japanese hunters on Wake had amassed some three-thousand bird skins, three months into a seven-month stint on the island.[60]

Matters played out differently on the atoll of Midway (Hawaiian: Kuaihelani). Midway had been annexed by the United States in 1859 as part of a disastrous attempt to establish a mid-Pacific coaling station. But since that failure no attempt had been made to settle it, and the preannexation Hawaiian government seems to have regarded it as its own territory.[61] However, in 1903 the Commercial Pacific Cable Company, with government backing, finally began laying the telegraph cable between San Francisco and Manila, and selected Midway (along with Honolulu and Guam) as one of several mid-Pacific landing stations.[62] In January 1903 Roosevelt issued an executive order placing Midway Island under the administration of the US Navy in order to 'guard against the invasion of Japanese sailors', as one newspaper put it. When the cable-landing party arrived at the island they discovered that a crew of Japanese bird-hunters had been established on the island

59 'The Marcus Island Case', *New York Times* (20 August 1902).
60 MOFA 3.5.2.8: Rear Admiral Evans to US Ambassador, Tokyo (6 January 1903).
61 NARA-II MD Territories Office, Hawai'i, Bird Reservations, Alexander to Dole (28 August 1900); Findlay, *Navigation of the North Pacific*, 1117; Jan Tenbruggencate, 'Historic Hawai'i Wreck Found', *Honolulu Advertiser* (18 October 2003). Last accessed 1 July 2015. For the Kingdom of Hawaii's claim to Midway, see HSA M-476 Pacific Chemical Fertilizer Co., 43.
62 Jonathan Reed Winkler, *Nexus: Strategic Communications and American Security in World War I* (Cambridge, MA: Harvard University Press, 2009), 146–7; 'Pacific Ocean Cable: Hearing Before the House Committee on Commerce', *Evening Star* (11 January 1902).

for months, and 'had practically denuded' the island of its albatrosses and terns.[63]

For two months the bird-hunters and the cable company staff resided together on the island uneasily, but when Captain Rodman of the USS *Iroquois* stopped at Midway two months later, he promptly ordered the Japanese off the island, citing the risk that the birds' decaying bodies might contaminate the island's water supply.[64] The following year the US Navy billeted an attachment of twenty-five naval marines to Midway, to protect the cable company staff but also to guard against poachers.[65]

It was encounters such as these that began to stir awareness in the US that Japanese bird-hunters were operating on the Northwest Hawaiian Islands. For instance, only a week after he had evicted the Japanese bird-hunters from Midway, Rodman discovered the same crew back at work on the neighbouring island of Lisianski (Hawaiian: Kapou). He once again ordered the bird-hunters to leave, and on his return to Hawai'i warned the Naval Department that 'for the past several years, the Japanese have been in the habit of visiting the various islands and slaughtering the birds in countless thousands'. In his report he provided a particularly harrowing description of the suffering that the Japanese hunters inflicted on the birds, that would become a staple of future accounts. According to Rodman the Japanese did not just butcher the birds, but did so in the most cruel ways, cutting off their wings while they were still alive, or leaving them crippled to starve to death.[66] (Other observers of bird-hunts described the killing as being done in a much more humane fashion than this, with another US naval officer reporting that only a very small percentage of birds were allowed to escape in wounded condition.)[67]

63 'Exterminated Midway Island', *The Hawaiian Star* (30 June 1903).

64 National Magazine Vol. XXI No.4 (January 1905) published by The Bostonian Publishing Co, Boston. Dr. Martin Cook, 'Our Cable Station in Mid-Pacific' (385–97), 388–9.

65 'Iroquois Ordered to Midway: To Reach the Island About Middle of May', *The Hawaiian Gazette* (20 March 1903).

66 NARA-II MD Territories Office, Hawai'i, Bird Reservations: Rodman to Assistant Secretary of Navy (1 July 1903).

67 'Thetis Visited Many Lone Islands' *Evening Bulletin* (23 June 1904). In this article other US naval officers furnished a rather sympathetic account of a Japanese bird-hunting operation they discovered, being 'of the opinion that the killing was carried on in a more humane manner. They found only one maimed bird. This bird had only one wing, the other having been cut off. This

Significantly, Rodman then went on to advocate expelling Japanese bird-hunters from *all* the islands of the Northwest Hawaiian group, widening the cable station's security perimeter from a single atoll to an entire archipelago, and eliding the distinction between national security and bird conservation. He found a willing ally in William Alanson Bryan, whose experience on Marcus Island two years earlier must have still been fresh in his memory. Bryan wrote to the State Department in support of Rodman's campaign for bird protection, and also enlisted the support of the American Ornithologists' Union, whose chairman wrote to the Naval Secretary begging him to 'establish rules and regulations as will prevent the killing and taking of resident birds [on Midway] for commercial purposes, and also to prevent the taking of the eggs of the same birds during the breeding season'.[68]

These combined missives precipitated a flurry of administrative activity, as different branches of government consulted each other as to what could be done to prevent the Japanese 'depredations'. The Navy ordered its 'commanding officers in Hawaiian waters . . . to give all the aid within their power with a view to suppress the destruction of the birds'. Governor Dole petitioned the State Department to issue a proclamation protecting the birds from the ravages of foreign nationals, or even from anyone who was not a resident of Hawai'i.[69] The State Department in turn consulted the Attorney General, warning that 'unless something is done to prevent such ravages, the fowl will in a few years be destroyed and a source of valuable guano deposits destroyed'.[70] It also suggested that Dole use his own power as governor to introduce legislation prohibiting bird-hunting on the islands. But Dole was reluctant to do this – perhaps

had evidently been done by one of the men in direct disobedience to the manager's orders. Only a very small percentage of birds were allowed to escape in wounded condition.'

68 HSA Carter Bird Depredations: Acting Secretary of Interior Department to Governor of Hawai'i (4 November 1904); NARA-II MD Territories Office, Hawai'i, Bird Reservations: William Dutcher to Naval Secretary (2 July 1903).

69 NARA-II MD Territories Office, Hawai'i, Bird Reservations: Acting Secretary of Interior to Secretary of State (23 July 1903).

70 NARA-II MD Territories Office, Hawai'i, Bird Reservations: Acting Secretary of State to Attorney General (28 July 1903).

because he knew how much it would cost the territorial government to enforce such a law.[71]

Instead, attention focused on whether the bird-hunters could be found to have violated existing federal laws. Rodman had suggested that the Japanese bird-hunters were in violation of US law, either immigration or customs, or perhaps quarantine law.[72] Dole made the same argument to the State Department.[73] Eventually the Treasury agreed to appoint Rodman 'Inspector of Customs for Midway and the outlying islands' so that, even if bird-hunting on the Northwest Hawaiian Islands was not technically illegal, he could prosecute Japanese hunters for violating customs, quarantine or immigration laws.[74]

Finally, there were diplomatic avenues to pursue. After the incident on Midway, the American ambassador to Tokyo argued that bird-hunting did not fall within the definition of 'peaceable occupations' as had been claimed after the incident on Wake Island. He explained that the destruction of birds by Japanese 'marauders' was resulting in 'unsanitary conditions' that were endangering the new cable station at Midway.[75] The ambassador's memorandum combined two separate arguments, an explicit hygienic one and an implicit conservationist one. The hygienic argument was somewhat specious, in that it may have justified evicting Japanese hunters from Midway, where the cable station was in operation, but hardly made sense when applied to otherwise uninhabited islands. The conservationist argument was hinted at only obliquely, through terms such as 'depredation' and 'marauder' which, like the adjective 'wanton' (another term often deployed by wild bird protectionists) convey moral disapproval of the killing of birds without pointing to any specific legal infraction. Foreign Minister Komura, for his part, responded to the memorandum

71 NARA-II MD Territories Office, Hawai'i, Bird Reservations: Acting Secretary of State to Interior Secretary (15 August 1903).

72 NARA-II MD Territories Office, Hawai'i, Bird Reservations: Rodman to Assistant Secretary of Navy (1 July 1903).

73 NARA-II MD Territories Office, Hawai'i, Bird Reservations: Acting Secretary of Interior to Secretary of State (23 July 1903).

74 NARA-II MD Territories Office, Hawai'i, Bird Reservations: Rear Admiral Silas Terry to Assistant Secretary of the Navy (18 November 1904).

75 MOFA 3.5.8.2 'Middouē'-jima kaichō hokaku kinshi 'Middouē'-jima oyobi sono hoka no Hawai guntō ni okeru honpō-jin no kaichō hokaku kinshi ikken: Griscom to Komura, [n.d.].

noncommittally, merely stating that 'the subject has been referred to the Authorities concerned'.[76]

The Lisianski Incident

In the months after the Midway Incident, reports continued to circulate of Japanese bird-hunters active in the Hawaiian islands. In November 1903, a Japanese schooner of about one hundred tons was sighted off the French Frigate Shoals (Hawaiian: Kānemiloha'i). Though the ship's captain claimed to be hunting sharks, the Navy believed this to be 'a mere pretext, and that the real object of this vessel's presence is the destruction of birds on these islands for their feathers'.[77] Dole's successor as Governor of Hawai'i went so far as to write to the captain warning him (inaccurately) that bird-hunting in the Northwest Hawaiian Islands was 'now a violation of federal law.'[78] The following April, the USS *Iroquois* returned to Lisianski to find seventy-seven Japanese men busily involved in the slaughter of birds. The *Iroquois* did not have room onboard to arrest such a large party, so Rodman notified them to desist.

Matters came to ahead two months later, in June 1904, when the revenue cutter *Thetis* returned to Lisianski to discover the cull still in full swing. This time, the hunters were brought aboard and taken to Honolulu, but under ambiguous circumstances: had they been arrested or rescued? According to the captain of the *Thetis*, the Japanese on Lisianski were on the verge of starvation, their ship having been wrecked on a reef five months earlier, and 'were very glad to leave the Island, as they had only 600 lbs. of rice and few beans left'.[79] But one newspaper cast doubt on this account, claiming that the hunters were not marooned but had been in contact with another ship, which had already arrived to collect a portion of the bird skins they had processed.[80] The Naval Department in its later correspondence would insist that the poachers had been arrested, not rescued.

76 MOFA 3.5.8.2 'Middouē'-jima kaichō hokaku kinshi: Komura to Griscom, [n.d.].

77 HSA Carter Bird Depredations: Terry to Carter (12 December 1903).

78 HSA Carter Bird Depredations: Carter to the Master of the Schooner 'Ada' (14 December 1903).

79 MOFA 3.5.8.2 'Middouē'-jima kaichō hokaku kinshi: Hamlet to Saitō (23 June 1904).

80 'Thetis Visited Many Lone Islands', *Evening Bulletin* (23 June 1904).

The problem was that there was no obvious charge to levy at the hunters, nor any legal basis for deporting them. Without determining by what (if any) contract they had been employed to hunt birds, they could not be shown to have violated immigration law. Indeed, most of the hunters soon signalled that they wanted to stay in Hawai'i, preferring to find work as canecutters rather than return to uncertain prospects back in Japan.[81]

Complicating things further was the matter of 200,000 processed bird skins that had not been brought back by the *Thetis* but left behind on Lisianski. The revenue cutter's captain had left a letter on the island, authorising the captain of the Japanese schooner that was scheduled to collect the hunters to 'take all their catch from the Island and carry it to Japan, with the understanding that no further bird killing should be done on any of the islands by Japanese hereafter'.[82] The question of who actually owned these bird skins quickly became a topic of hot discussion in the Hawaiian press. 'From a commercial standpoint,' noted one paper, 'the skins belong to any free lance that may choose to go after them.'[83]

The controversy quickly ensnared the Japanese consul to Honolulu. Consul Saitō Miki had a pivotal, but unenviable, diplomatic role in the Territory of Hawaii: charged with defending the rights of Japanese nationals to live and work in Hawai'i, but also with policing the behaviour of said nationals so as to avoid causing international embarrassment to the Japanese government. After the Midway Incident he had written to the Japanese Foreign Ministry, warning of the 'unusually furious' coverage that Hawaiian newspapers were devoting to the 'cruel behaviour' of Japanese hunters, and suggesting that the ministry take anticipatory measures to regulate the hunters and forestall future incidents.[84] Now he felt obliged to defend the interests of those same hunters, lodging a request on their behalf that the US government protect the bird skins on

[81] MOFA 3.5.8.2 'Middouē'-jima kaichō hokaku kinshi: Bechtel to Saitō (24 June 1904).
[82] MOFA 3.5.8.2 'Middouē'-jima kaichō hokaku kinshi: Hamlet to Consul Saitō (23 June 1904).
[83] 'Who Owns Those Rich Bird Skins on Lisianski?' *The Hawaiian Star* (24 June 1904).
[84] MOFA 3.5.8.2 'Middouē'-jima kaichō hokaku kinshi: Consul Saitō to Foreign Minister Komura (30 June 1903).

Lisianski 'from the interference of outsiders'.[85] (It later emerged that another Japanese vessel had indeed returned to Lisianski to collect the skins as scheduled.)[86]

At the same time, the Hawaii Inspector of Customs was keen that the Japanese consulate provide compensation for the expense of 'transporting' (the term was deliberately ambiguous) the rescued/arrested Japanese hunters from Lisianski to Honolulu. He requested that the Japanese government pay the immigration tax for the hunters who had chosen to stay in Hawai'i, and also bear the cost of repatriating the three men whose immigration petitions had been rejected, plus any other hunters who wished to return home.[87]

In effect, this amounted to an official request that the Japanese government accept liability for the actions of its nationals on uninhabited US territory, by shouldering the cost of expelling them. Consul Saitō refused to pay the transportation costs directly, but he did promise to 'endeavour to arrange for their passage at their own expense'. In practice this entailed a cross-governmental effort to track down the men's employer, a Tokyo-based fishery operator named Takahashi Fukujirō, who was eventually located by the Yokohama Port Police and pressured to pay the costs out of his own pocket.[88] The fallout from the Lisianski Incident therefore caused the Japanese government to become directly involved in managing the 'problem' of bird-hunting on the Northwest Hawaiian Islands. As Saitō noted in an internal memo, bird-hunting on the islands might not be technically illegal, but it risked inflaming anti-Japanese nativist sentiment that might lead to new restrictions on Japanese immigration more broadly. Better therefore to cooperate in suppressing the bird-hunts in the hope of averting future diplomatic blowback.[89]

85 'Thetis Visited Many Lone Islands'; HSA Carter Bird Depredations, Terry to Asst. Navy Secretary (18 November 1904); MOFA 3.5.8.2 'Middouē'-jima kaichō hokaku kinshi: Stackable to Saitō (20 July 1904).

86 'Iroquois Gets Back', *The Hawaiian Star* (29 September 1905).

87 MOFA 3.5.8.2 'Middouē'-jima kaichō hokaku kinshi: Customs Inspector Bechtel to Saitō (24 June 1904).

88 MOFA 3.5.8.2 'Middouē'-jima kaichō hokaku kinshi: Saitō to Bechtel (27 June 1904); Saitō to Foreign Minister Komura (26 September 1904); Saitō to Takahashi (22 September 1904); Takahashi & Nakamura to Saitō (30 August 1904); Acting Inspector of Customs Cullen to Saitō (6 October 1904).

89 Hiraoka, *Ahōdori to 'Teikoku' Nihon no kakudai*, 128.

Over the course of the next year the US government's attitude to bird-hunting hardened, and it became determined that the Lisianski Incident should serve as a line in the sand. Much of the credit for this shift in policy can be laid at the door of William Alanson Bryan, who wrote to President Roosevelt calling for the federal government to 'prevent the destruction of birdlife on the islands in Hawaiian waters'.[90] Bryan's memorandum had an immediate effect. The Treasury promised to arrange for the dispatch of another revenue cutter to patrol the waters of the Northwest Hawaiian Islands on an intermittent basis.[91]

Copies of Bryan's report were also sent to several government departments, including the Navy, the Department of Agriculture (which was responsible for management of wild game stocks) and the State Department, which now explicitly requested the Japanese government to help enforce a hunting ban.[92] In point of fact there was still no law on the books that forebade bird-hunting on the Northwest Hawaiian Islands, either at the Hawaiian or at the federal level. But the State Department now began referring to Japanese bird-hunters as 'poachers' – language that found its way into Japanese government reports as well.[93] It even supplied a list of firms believed to have engaged in poaching activities and extracted an assurance that police forces in Tokyo, Yokohama and Taiwan would force the suspects to comply with the ban.[94]

The Lisianski Incident therefore helped form a new diplomatic consensus, whereby the Japanese government committed to taking pre-emptive measures to suppress the activities of bird-hunters on the Northwest Hawaiian Islands. This chimed with a broader shift in US–Japan borderland policy. Over the course of the 1900s the Japanese government implemented increasingly stringent screening of passport applications, proactively restricting labour emigration to the continental US to avoid being targeted by an explicit immigration ban – a process that culminated in the 1907 Gentlemen's

90 HSA Carter Bird Depredations: Acting Treasury Secretary to Interior Secretary (2 November 1904).

91 HSA Carter Bird Depredations: Taylor to Interior Secretary (2 November 1904).

92 HSA Carter Bird Depredations: Acting Secretary of State to Interior Secretary (3 November 1904).

93 MOFA 3.5.8.2 'Middouē'-jima kaichō hokaku kinshi: Sugihara to Foreign Minister Saionji (5 April 1906).

94 HSA Carter Bird Depredations: Komura to Griscom (21 February 1905).

Agreement.[95] And in 1911 Japan, the US, Great Britain and Russia all signed the North Pacific Fur Seal Convention, a landmark treaty in which signatories banned their citizens from engaging in pelagic fur sealing north of the 30th parallel. Significantly, signatories also committed to preventing their own nationals from poaching onshore Alaska's seal-rich (but sparsely populated) Pribilof Islands.[96] In all these cases the US successfully pressured Japan to not only affirm its territorial sovereignty claims but also actively curb the trans-Pacific mobility of Japanese subjects.

Nature and Labour on Laysan Island, 1890–1900

For the next few years there was a lull, as reports of Japanese bird 'poaching' in the Northwest Hawaiian Islands died down. Perhaps this was because the new raft of policies succeeded in deterring future bird-hunting expeditions. Or perhaps it is simply that after 1905 bird populations on islands such as Lisianski and Wake were already so depleted that plumage companies no longer thought it commercially worthwhile to harvest them. Either way, at the end of the decade the issue flared up again dramatically. This time the dispute swirled around the island of Laysan (Hawaiian: Kamole),[97] the only one of the Northwest Hawaiian group that had been inhabited prior to Hawaiian annexation.

The Laysan Island Incident deserves sustained attention for a number of reasons. For one thing the island's history represents an almost perfect microcosm of labour and race relations in post-annexation Hawai'i. It also represents a culmination of several trends discussed previously: increasing white anxiety about the activity of Japanese bird-hunters in nominally American sovereign space, and coordinated diplomatic and coercive efforts to expel said bird-hunters from that space. The incident also led directly the establishment of the

95 Mitziko Sawada, 'Culprits and Gentlemen: Meiji Japan's Restrictions of Emigrants to the United States, 1891-1909', *Pacific Historical Review* 60 no. 3 (August 1991): 341.

96 Scott Barrett, *Environment and Statecraft: The Strategy of Environmental Treaty-Making* (Oxford: Oxford University Press, 2005), 42. According to the 1910 US census only two of the four Pribilof Islands were inhabited, with a combined population of under 300. See US Census Bureau, *Supplement for Alaska* (Washington, DC: Government Printing Office, 1913), 574.

97 Kikiloi, 'Kūkulu Manamana', 49.

Hawaiian Islands Reserve, which set a precedent for a number of other nature reserves within the bounds of the United States' Pacific empire and laid the foundation for a later generation of expansive marine national parks, including today's Papahānaumokuākea Marine National Monument. In this sense its ramifications stretched far beyond the confines of Hawai'i, far along the shoreline of Alaska and deep into the Pacific Ocean.

We have a good description of Laysan written by schooner captain John Paty, who landed on the island in 1857 and claimed it for King Kamehameha IV of Hawai'i. Paty described Laysan as a 'low sand island, 25 to 30 feet high', its surfaces 'covered with beach grass' and 'literally covered' with birds, 'evidently unaccustomed to the sight of man, as they scarcely move at our approach, and the birds are so tame and plentiful, that it was difficult to walk about the island without stepping upon them ...'[98] Paty also mentioned the island's most unusual feature for a North Pacific atoll, its central lagoon with an 'abundance of tolerable good fresh water'. Because of this it was home not only to seabirds, but also to various species of freshwater birds whose rarity and evolutionary distinctiveness would bring ornithologists flocking to the island when it was first brought to the attention of the wider world.

In 1890, an Englishman named Freeth rented Laysan from the Hawaiian government for $5 per year and went on to open a guano mine on the island. This mine on Laysan owed its existence to the remarkable sugar boom that had transformed Hawai'i over the previous decades. Freeth's company, the newly formed Pacific Guano & Fertilizer Company, received majority financing from Hackfeld & Co., a German planters' firm based in Kaua'i, and Hackfeld would use the guano it produced to fertilise the cane-fields on its plantations.[99] Like those sugar plantations, the Laysan mine relied for its operation on indentured Japanese labourers (Figure 3.2), an early attempt to recruit workers from the Gilbert Islands (Figure 2.1) having turned sour.[100]

98 Paty, 'Account of the Manuokawai – Interesting Account of Her Explorations'.

99 The company was initially named the North Pacific Phosphate and Fertilizer Company, and was renamed in 1895. See HSA M-476 Pacific Chemical Fertilizer Co.

100 HSA M-476 Pacific Chemical Fertilizer Co: Minutes (2 April 1891). On the early presence of I-Kiribati labourers, see Tom Unger, *Max Schlemmer, Hawai'i's King of Laysan Island* (Bloomington: iUniverse, 2004), 21.

Figure 3.2 Japanese guano miners on Laysan (c.1903)
Laysan albatrosses (*Phoebastria immutabilis*) are visible in the background
Source: University of Hawai'i at Mānoa William Alanson Bryan collection

And, again like the sugar plantations, the Laysan mine experienced an outbreak of industrial unrest immediately after US annexation and the abolition of indentured labour in Hawai'i – one that ended with two Japanese labourers shot dead and their white overseer arraigned in the Honolulu District Court on charges of manslaughter.

'Laysan Island's Story of Blood', as *The Hawaiian Gazette* luridly dubbed it, crystallised white anxieties about race and labour relations in post-annexation Hawai'i, with the tiny island operating as a synecdoche for the whole archipelago in the fraught summer of 1900 following the passage of the Organic Act.[101] Accounts differ as to what chain of events led to the shootings. At the arraignment, two Japanese testified that the dispute began when they had asked for an altered work schedule to accommodate the fact that 'the guano dust, blown by the fresh sea breeze, got into the eyes of the laborers and greatly bothered them'. When this was refused the miners downed tools

[101] 'Laysan Island's Story of Blood', *The Hawaiian Gazette* (11 September 1900).

in protest, in response to which the overseer, a septuagenarian known as Captain Spencer, refused to sell rice to the miners unless they resumed work.[102] Based on the Japanese testimony the sheriff (reluctantly) sought manslaughter on the basis that Spencer had 'attempted to starve the men into submission . . . it would have been inhuman to cut off food, even if they did refuse to work, being on a barren island as they were.'

The defence counsel, however, claimed that the Japanese had engaged in threatening behaviour, such as wrestling on the beach, banging drums and raising the Japanese national flag above their house.[103] Matters culminated in the miners surrounding Spencer's house 'armed with knives, clubs stones and cutlasses made of hoop-iron sharpened' (the Japanese denied carrying any weapons), 'cursing the ageing captain and his friends and shouting something to the effect that they were American citizens [a misinterpretation of the Organic Act] and wouldn't work any more on the Island but would behave thereafter as best suited themselves'.[104] Spencer and his assistant drew their firearms in warning, but when the miners rushed them (so the defence argued) they had no choice but to fire, for 'when a man goes up against a loaded revolver he is out for blood'.[105]

Local press coverage was overwhelmingly sympathetic to Spencer. For *The Hawaiian Gazatte*, 'War has been declared, waged and ended on Laysan Island . . . The forty Japanese rose in a body, determined to annihilate all the white people on the Island and run things to suit themselves . . . They threatened the white men's lives . . . . [But] though they moved quickly, Captain Spencer's trigger fingers moved quicker.' The Hawaiian-language (but pro-American) *Ka Nupepa Kuokoa* also took Spencer's side, describing him as a 'very distinguished resident of Hawai'i (kama'āina)' and stressing that his crime 'arose out of aversion to the damage that the Japanese intended to inflict on one of the few white settlements (wahi haole) in the Northwest Hawaiian Islands'.[106] After only an hour of argument the magistrate dismissed the case,

102 'White Men All Fired, Say Laysan Island Japs', *The Hawaiian Star* (14 September 1900).
103 *Ibid.* 104 'Laysan Island's Story of Blood'.
105 'Captain Spencer is Promptly Dismissed: Laysan Island Case Thrown Out of Court', *Honolulu Republican* (22 September 1900).
106 'Kipu ma ka mokupuni lepo manu', *Ka Nupepa Kuokoa* 38 no. 37 (14 September 1900).

declaring that 'it would be impossible to find a jury that would convict the captain of any crime'.[107]

The United States vs. The King of Laysan

After this debacle, the Pacific Guano & Fertilizer Company decided to look for new management to run the mine on Laysan. They found their man in one Max Schlemmer. Schlemmer was, like Rosehill, a latter-day beachcomber, another of those Melvillian characters who were to be found drifting round the global nineteenth-century Pacific, ekeing out a living on the shorelines of state sovereignty. Though ethnically German, Schlemmer was in fact born in French-administered Alsace. In 1871, the year that Prussia invaded and annexed his home province, he left home, bound for the United States, aged fifteen. Within a few months of arriving in America he had found work on a whaler out of New England, and spent the greater part of the next two decades criss-crossing the Pacific before eventually washing up on Kaua'i in Hawai'i to become a fixture of the island's small German community.[108]

In 1890, when the Pacific Guano & Fertilizer Co. announced its plan to begin mining guano on Laysan, Schlemmer had approached the firm's board of directors claiming to have lived on the island for a stint five years previously. On the basis of this claim to squatters' rights, he succeeded in negotiating a small royalty from the mine's guano sales. (Whether or not Schlemmer had actually set foot on Laysan before that point is unclear.)[109] In 1893 he agreed to take a job as foreman at the Laysan mine, and spent the next six years supervising the Japanese labourers under the authority of the overseer Spencer. In 1899 he quarrelled with Spencer and decided to return to Honolulu. So when Spencer's tenure as overseer ended in disaster the following year, Schlemmer was an obvious candidate to replace him.

The English-language Hawaiian press was fascinated by Schlemmer. There was the fact that he lived a Swiss Family Robinson-like existence, accompanied on Laysan by his wife and eleven children (Figure 3.1c). And he was known as a genial host, accommodating visiting naval and scientific expeditions, including both Rodman and Bryan, and

107 'Shooting Was Justified', *The Hawaiian Star* (22 September 1900).
108 Unger, *Max Schlemmer*, 1–3.
109 Star Bulletin Printing House, *Sales Builder* (Honolulu, January 1939), cited in Unger, *Max Schlemmer*, 9–10.

furnishing them with his extensive knowledge of the island's flora and fauna – its birdlife in particular.[110] They soon took to referring him as the 'King of Laysan'.[111]

Of course, it was meant half in jest. Surely, by the turn of the twentieth century, every right-thinking citizen of the United States knew that kingship had been consigned to the dustbin of history? The white community in Hawai'i was particularly sure of this – had not the Kingdom of Hawai'i, just a few years earlier, been absorbed peacefully by the civilising power of the Republic? Then again, the name was only partly a joke. The inhabitants of the scattered Hawaiian archipelago understood better than most that remoteness confers a high degree of autonomy. The precedent of the 1900 shootings also suggested the use of force on Laysan Island would likely go unpunished. *The Hawaiian Star* described Schlemmer as 'the supreme being of his little kingdom. His Japanese workmen must obey his command, lest they be deported or, under rarer circumstances, shot to death, which in the latter event is looked upon as a justifiable act, although the Laysan monarch is under the indirect control of the United States.'[112]

Schlemmer for his part relished his royal status, but also made every effort to accentuate his American-ness. Unlike Rosehill, another European immigrant to Hawai'i, Schlemmer had fought to defend the Hawaiian monarchy against pro-US annexationist forces.[113] But after 1898 he moved quickly to adopt US citizenship, took care to fly the Stars and Stripes on Laysan whenever ships sailed past, and generally emphasised his credentials as 'an American citizen who has always been a friend to Navy and Army people and also to scientific people all over the Globe . . .'[114]

In April 1904 Hackfeld and Co. decided to divest from the guano mine in favour of the emerging sector of chemical fertilisers.[115] So

110 'Princess of Laysan and Olson's Disaster'; 'Exterminated Midway Island'.

111 'The King of Laysan: His Majesty Here for a Short Stay', *The Hawaiian Star* (29 July 1901). In fact, he had already conferred this title on his predecessor, Captain Spencer, before his ill-fated reign was cut short.

112 'Princess of Laysan and Olson's Disaster'.

113 Unger, *Max Schlemmer*, 16–17.

114 NARA-II MD, Territories Office, Hawai'i, Bird Reservations: Schlemmer to Kalaniana'ole, [undated].

115 HSA M-476 Pacific Chemical Fertilizer Co. In the years 1896–1899, the company extracted on average 2,565 lbs of guano for every dollar spent on labour. For the period 1901–1903 this figure dropped to 1,671 lbs.

when Schlemmer expressed an interest in buying the remaining six years of the Laysan leasehold, the Pacific Guano & Fertilizer Co.'s board was only too happy to sell it to him.[116] It seems likely that, from the outset, Schlemmer planned to supplement the income to be gleaned from guano mining by expanding into the plumage trade. Soon after acquiring the lease he petitioned the Governor of Hawai'i to extend his leasehold for Laysan to ninety-nine years. His petition showed a shrewd understanding of the way the political winds were shifting. Mindful of the increasingly influential bird conservationist lobby, he promised to protect the islands' birdlife from Japanese poachers. In return he asked permission to harvest a sustainable number (he gave the oddly specific figure of 21,800 per season), which he would turn over to the territorial government to sell.[117]

Schlemmer was not the only Honolulu-based entrepreneur hoping to get into the plumage business during this period. After the Lisianski Incident in the summer of 1904 a businessmen named Atkinson openly held talks with Sugie Tsunetarō, the foreman of the arrested/rescued Japanese hunters. Atkinson was a man of some political clout, having briefly served as Acting Governor of Hawai'i. 'There is no reason', he argued, 'why the territory should not lease the bird killing privileges to such persons as have the enterprise to go there. The birds would of course be protected from extermination and that at present seems to be the only fate in store for them.'[118]

Like Schlemmer, Atkinson was deploying a kind of proto-Hardinian logic: that granting property rights over the birds would incentivise the owner to harvest them sustainably, thereby averting the 'tragedy of the commons'.[119] Indeed, at one point Schlemmer even considered forming a consortium with Atkinson and Sugie, but the men eventually fell out bitterly. (He would later accuse them of conspiring with his own guano

116 MOFA 3.5.8.2 'Middouē'-jima kaichō hokaku kinshi, 361.

117 HSA: Carter to Schlemmer (23 December 1904).

118 'To Lease Lisiansky', *Hawaiian Star,* 29 June 1904.

119 Garrett Hardin, 'The Tragedy of the Commons', *Science* 162 no. 3859 (13 December 1968). It should be said that environmental economists have long since debunked Hardin's theory. Colin W. Clark constructing a model based on Antarctic whale stocks which showed that 'depending on certain easily stated biological and economic conditions, extermination of the entire population may appear as the most attractive policy, even to an individual resource owner'. Colin W. Clark, 'Profit Maximization and the Extinction of Animal Species', *Journal of Political Economy* 81 no. 4 (1973).

miners to poison his family and pillage Laysan's birds.)[120] Such shifting alliances do make it clear, however, that no matter who owned the rights to Laysan's birds, the actual work of killing them would most likely be done by Japanese.

Governor Carter, Sanford Dole's successor, was unenthusiastic about such schemes. He agreed to forward Schlemmer's leasehold request to Washington, DC, but rejected outright the suggestion that the administration itself get involved in the plumage trade.[121] But Schlemmer persisted, reaching out to potential allies among the naval officers and naturalists (including William Alanson Bryan) who had visited Laysan during his tenure there. He also wrote to Prince Kalaniana'ole, Hawai'i's elected delegate to Congress, asking for help securing a federal appointment as police warden to protect the Northwest Hawaiian Islands from poachers. Like Schlemmer, Kalaniana'ole was a former anti-Annexationist who had recently accommodated himself to US rule over Hawai'i – by joining Roosevelt's Republican Party, no less.[122] Perhaps hoping to remind the prince of his prior support for the Hawaiian monarchy, Schlemmer signed off his letter in Hawaiian.[123]

As a result of this lobbying, in 1907 Schlemmer did receive an unpaid commission as a police constable with responsibility for patrolling the Northwest Hawaiian Islands. This in no sense constituted a license to hunt birds; indeed, the commission explicitly stated that the authority granted to him was primarily for the *protection* of birds. But by this point Schlemmer's finances were becoming increasingly precarious, and only three years remained on his lease to Laysan. The following April he announced that he was to construct a tannery on Laysan, ostensibly for the purpose of processing the skins of the rabbits he

120 HSA Frear Bird Depredations: Schlemmer Testimony (8 July 1908).

121 HSA: Carter to Schlemmer (23 December 1904).

122 Noenoe K. Silva, 'The First Election of Ke Keikiali'I Kūhiō Kalaniana'ole', in *Biography Hawai'i: Five Lives* (Honolulu: University of Hawai'i at Mānoa Center for Biographical Research, n.d.): 4.

123 NARA-II MD, Territories Office, Hawai'i, Bird Reservations: Schlemmer to Kalaniana'ole, [undated]. There is no record of Kalaniana'ole's reply to Schlemmer, nor of him voicing any opinion on the fate of the Northwest Hawaiian Islands during this period. He did, however, support the territorial government's plan to promote white immigration from Europe 'in order to get rid of the Asiatics in our country'. See 'Col Hepburn's Presence Needed: His Illness Kept Him Out of Immigration Debate', *Pacific Commercial Advertiser* (9 May 1908).

had released to breed on the island.[124] Then in December 1908 he made his boldest play yet. He travelled to Tokyo, where he signed a fifteen-year contract with one Yamanouchi Genkichi, owner of a deep-sea fishing company, granting Yamanouchi 'the privilege of securing phosphate guano, and any products of whatever nature in and from [Laysan Island]'.[125] Schlemmer also undertook to use his police constable's commission for the purpose of 'overseeing any other people who would come to the said Islands and impinge [Yamanouchi's] privilege' – a highly flexible interpretation of his original mandate – in return for $150 per month.[126]

Schlemmer's gambit, then, was to try and leverage the authority vested in him as police warden to effectively sublet Laysan Island to the very bird-hunters he was charged with repelling. How he thought he could pull this off is unclear. The deal he struck was, in theory at least, not too far removed from the various schemes that had floated around Honolulu for years by this point: to set up some kind of government-authorised, notionally sustainable scheme for harvesting Laysan's seabirds. He seems to have still hoped that the Hawaiian territorial government might grant him his cherished leasehold extension, which would perhaps help legitimise the legally dubious arrangement he had struck up. Perhaps he was hoping that nobody would notice, or that nobody would care.

If so, he miscalculated. News of Schlemmer's trip to Japan soon circulated, triggering outrage in first the Hawaiian and then the mainland US media. It made the pages of both the *San Francisco Chronicle* and the *New York Times,* from whence it caught the eye of William Dutcher, then president of the National Union of Audubon Societies. Dutcher promptly sent the *Times* clipping to his old colleague William Alanson Bryan. Bryan, who happened to be in Washington, DC at the time, in turn forwarded Dutcher's letter to the State Department, hand-delivering it to Foggy Bottom himself.[127] This set off a panic within the federal government, compounded by

124 Unger, *Max Schlemmer*, 64, 136.

125 MOFA 3.5.8.2 'Middouē'-jima kaichō hokaku kinshi: Max Schlemmer, 'License' (22 December 1908).

126 MOFA 3.5.8.2 'Middouē'-jima kaichō hokaku kinshi: Schlemmer to Yamanouchi (22 December 1908).

127 HSA Frear Bird Depredations, State Department Chief Clerk to Bryan (13 January 1909).

uncertainty as to whether Schlemmer's contract had any legal basis and, if so, what steps could be taken to overrule it. On 3 February 1909, shortly before Theodore Roosevelt left office, the Interior Secretary submitted a memorandum suggesting that the President use his executive power to declare the Hawaiian islands a refuge for 'native birds ... reserving them for public use'. Roosevelt signed an executive order the same day, creating the Hawaiian Islands Reservation (Figure 3.3).[128]

Over the following year, newspapers began reporting that Japanese vessels had resumed their visits to the Hawaiian islands. Yamanouchi, it turned out, had decided to subcontract the rights granted him by Schlemmer to another fishing company, and eventually the hunt for Laysan's birds devolved into a free-for-all.[129] One Tokyo newspaper reported that at least seven different ships belonging to a variety of different operators had visited Laysan during the winter bird-hunting season of October 1908 to January 1909: 'coolies are crowded on this small island as thickly as spectators at a fire ... it is feared that business rivalry may bring them to quarrelling'.[130] (An alternative possibility is that Yamanouchi, rightly doubtful that his contract with Schlemmer would prove enforceable in the long-term, came to an arrangement with other bird-hunting companies to jointly harvest Laysan's bird population as swiftly as possible.)

Reports of renewed bird depredations reached the American authorities as well. In March 1909 the cable station superintendent reported that a Japanese schooner stopped at Midway for general repairs, likely bound for Laysan and other uninhabited atolls in the Northwest Hawaiian group.[131] By the end of the year, reports of 'Jap poachers ... stripping the small islands between here and Midway of

128 NARA-II MD Territories Office, Hawai'i, Bird Reservations: Dennett to Interior Secretary (3rd February 1909); Executive Order 1019, viewable at: www.presidency.ucsb.edu/documents/app-categories/written-presidential-orders/presidential/executive-orders. Last accessed 9 February 2022.

129 Japanese: *Nihon Gyogyō Kabushiki Gaisha*. See MOFA Middouēi: Hatoyama to Ministry of Agriculture and Commerce Marine Products Department, 28 February 1910; Governor of Kanagawa to Foreign Minister Komura, 17 March 1910.

130 *Chūgai Shimpō* 'Expedition of Japanese Sailing Vessels to Secure Feathers', 29 March 1909 cited in NARA-II MD Territories Office, Hawai'i, Bird Reservations: Poaching of Japanese on bird islands of Hawaiian group, 3 April 1909.

131 G. V. Meyer to unknown, 31 March 1909.

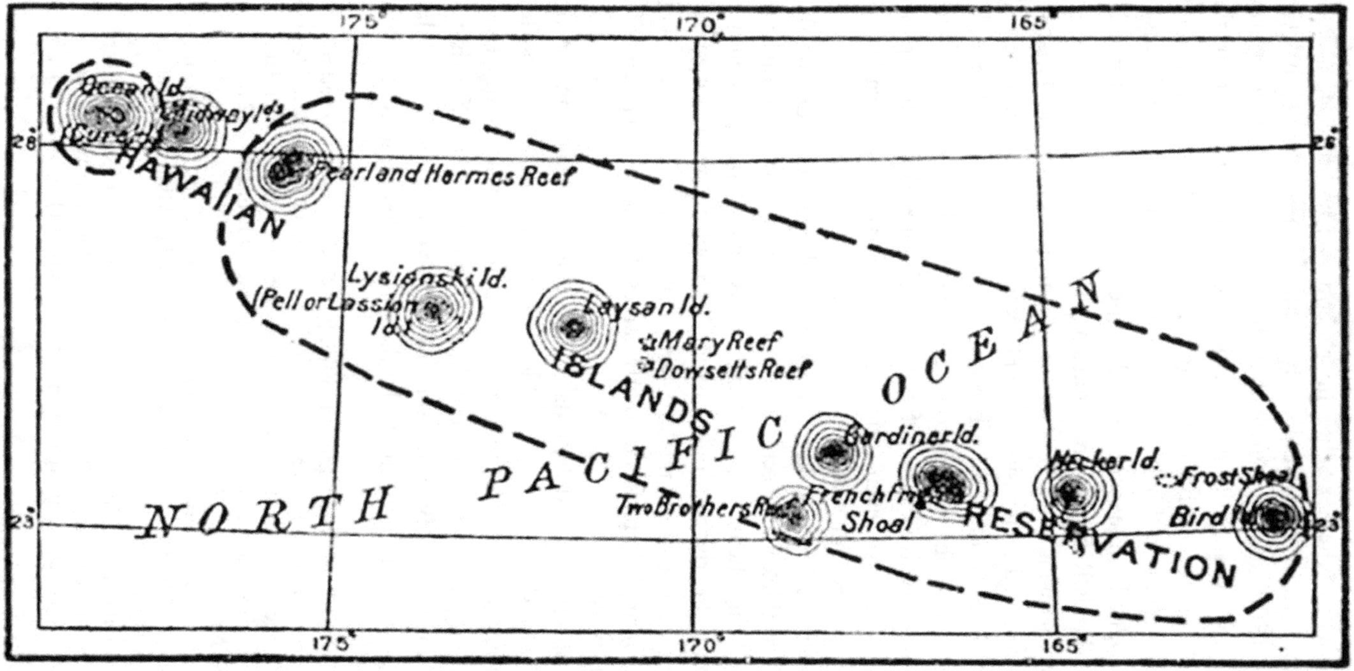

Figure 3.3 The Hawaiian Islands Reservation (1909)
Source: en.wikisource.org/wiki/Executive_Order_1019

birds' were again appearing regularly in the Hawaiian press.[132] A reader who claimed to be familiar with the islands framed the problem in nakedly inter-imperial terms, condemning:

> what I see to be a disgrace to a civilized power, in allowing Japanese year after year to destroy the ocean sea birds ... It is all very well issuing a proclamation that the islands above mentioned are the sole property of Uncle Sam and no one must kill any birds, etc etc. Bah! ... Just imagine for a moment Kaiser William or Edward the Seventh permitting such atrocities to be continued year after year on their domains – but don't imagine, it is too absurd for imagination![133]

The reports triggered a renewed effort to crack down on bird-hunters. At the behest of the State Department the Japanese Foreign Ministry placed newspaper notices advertising the establishment of the Hawaiian Islands Reservation, including in the Honolulu-based *Hawai Shimpō*.[134] And the Home Ministry instructed coastal authorities to warn Japanese ships of a heightened risk of arrest for poaching in Hawaiian territory.[135] The US government also took more direct steps to suppress the hunting, scrambling to requisition its single Pacific revenue cutter, the *Thetis,* at the time deployed in Alaskan waters.[136]

In January 1910 the *Thetis* stopped at Laysan Island to discover twenty-three Japanese engaged in a systematic bird-slaughtering and skinning operation. The captain arrested the hunters for poaching, and transported them to Honolulu along with over a quarter of a million bird wings. It also confiscated the Schlemmer-Yamanouchi contract, presented by the hunters' foreman as proof that they were operating on the island legitimately. This damning piece of evidence made it clear who the real villain was. As *The Hawaiian Star* intoned darkly that the captured Japanese were merely the pawns of nefarious commercial interests: 'It may go hard with them for what they've done, but it

132 HSA Frear Bird Depredations: Stackable to Frear, 7 October 1909; *Hawaiian Gazette*, 'Jap Poachers on Islands to Windward', 1 December 1909.

133 'Slaughtering the Sea Birds', *The Honolulu Sunday Advertiser* (24 October 1909).

134 MOFA Middouē: MOFA Commerce Dept to Assistant Foreign Minister (27 January 1910); MOFA to US Ambassador (26 January 1910); 'Dōhō gyogyōsha e chūi', *Hawai Shimpō* (26 October 1909).

135 MOFA Middouē: 'Hawai guntō ni okeru kaitō mitsuryō ni kan suru ken', 467.

136 NARA-II MD, Territories Office, Hawai'i, Bird Reservations: Acting Treasury Secretary to Interior Secretary (23 July 1909).

appears that they were but the tools of a great company, whose headquarters is in Japan and which has agents in Hawai'i.' The real culprit, though, was Schlemmer: 'A Honolulu man, a white man, is supposed to know all about their business and he will have much to explain if not to answer for.'[137]

The US government was also determined to make an example of the King of Laysan. A federal grand jury arraigned him on two counts of poaching from a federal reservation and bringing aliens into the country unlawfully. But the prosecution soon found itself bogged down in a legal morass. Schlemmer's defence lawyer argued that Roosevelt's order authorising the creation of the Hawaiian Islands Reservation rested on dubious constitutional grounds. (A government memo also admitted that 'Congress has since attacked the proposition that the President has the power to reserve public lands for such purposes'.[138]) Neither could Schlemmer be prosecuted under the laws of the Territory of Hawaii, as there was no specific anti-poaching law on the books that covered the Northwest Hawaiian Islands. Eventually, the prosecution failed to convict Schlemmer either on the charge of poaching or on illegal employment of foreign labour. Interestingly, the Honolulu district court judge accepted the defence's argument that the Hawaiian Islands Reservation, even if constitutional, only provided protection for 'native birds'; whereas the prosecution had failed to provide any evidence as to whether the birds killed on Laysan were native or not. Formally at least, Schlemmer was exonerated.

Schlemmer for his part was utterly unrepentant. He claimed, somewhat speciously, that the Japanese on Laysan were guano miners under his employ who merely 'killed a few birds on the side'. He also claimed that he was being persecuted by the US government merely for selling his rights to Laysan to a Japanese national.[139] This claim has the ring of truth to it. After all, ever since the first reports of bird culls on otherwise uninhabited Hawaiian islands, the US government had been exercised by the fact the 'depredations' occurred specifically at the hands of Japanese nationals. The panic about Japanese bird

137 'Poachers Brought by Thetis: Max Schlemmer Thought to Be in Poaching Deal', *The Hawaiian Star* (2 February 1910).

138 NARA-II MD, Territories Office, Hawai'i, Bird Reservations: 'Hawaiian Islands Reservation – Memorandum' (n.d.)

139 'Schlemmer Says Fear Cause of Trouble', *Honolulu Evening Bulletin* (3 February 1910).

depredation also occurred against a background of widespread anxiety about Japanese migration into United States territory and encroachment into economic activities that were properly felt to be the preserve of US citizens, anxiety that had crystallised so vividly in William Alanson Bryan's experience on Marcus Island in the summer of 1902. In this sense, the crackdown on Japanese bird-hunting was indeed part of a broader effort to exclude Japanese from the political and economic life of the Territory of Hawaii.[140]

The State of Nature

The fallout from the Laysan Island Incident occurred on multiple scales. For Schlemmer, his acquittal was a pyrrhic victory: the trial had exhausted his finances, and although the Territory of Hawaii had in the end agreed to extend his lease on Laysan, this was now annulled in light of Roosevelt's executive order. One newspaper wryly noted the contrast with the murder arraignment of Captain Spencer nine years earlier: 'There has been no revolt among the subjects of King Max. His rule has been mild and beneficent and his people have always been well-satisfied. It is only a case of a stronger power stepping in and taking away his land. The US government is the power that has deposed the King of Laysan and annexed his country.'[141]

For Bryan and his conservationist allies the victory was a mixed one. In the aftermath of the trial Bryan participated in a Department of Agriculture expedition to the Northwest Hawaiian Islands – 'Uncle Sam's great ocean bird reserve', as one newspaper put it. The expedition aimed first of all to survey the islands' wildlife, partly to determine which of its fauna could be classified as 'native'. Given the fiasco of the Schlemmer trial, this was a matter of pressing legal importance if future poachers were to be successfully prosecuted. Secondly, the USDA expedition tried to help Laysan's ecology recover from the impact of

[140] To this should be added the fact of Schlemmer's German-ness, which may not have rendered him as irredeemably alien in the same way as the Japanese, but nevertheless marked him as foreign in the new US colony. During World War I he would even be briefly detained on suspicion of spying on American shipping. There was some irony in this accusation, as few people had demonstrated their distaste for German militarism more manifestly than Max Schlemmer, who had left his homeland of Alsace at the very moment it was invaded by Prussia. See Unger, *Max Schlemmer*, 85–6.

[141] 'King Max of Laysan is Deposed' *Honolulu Advertiser* (11 November 1909).

Japanese hunting, but also of human occupation more generally. During his time on the island, Schlemmer had undertaken various initiatives to modify his domain for human habitation, planting coconut palms and releasing pigs and rabbits to breed on the island as a food source.[142] But the USDA scientists deemed these new introductions a menace to Laysan's 'native' bird population, and so a new cull commenced on Laysan, this time of the island's 'invasive' species. Unfortunately, the expedition exhausted its supplies of shotgun ammunition before it was able to wipe out the last remaining rabbits, whose population soon rebounded.[143]

Despite this failure, Bryan did succeed in restoring Laysan to its wild, prelapsarian state – at least in simulacrum. Together with his old adviser at the University of Iowa he raised funds to install a three hundred and sixty-degree diorama in the university's natural history museum. Visitors to the museum can duck their heads into a darkened room and surround themselves with an artfully composed tableau of stuffed bird carcasses, merging imperceptibly into a painted backdrop depicting island and ocean (Figure 3.4).[144] Much like the displays installed at the American Museum of Natural History by the great Carl Akeley, the diorama was designed to suggest unhindered access to raw nature in its pristine state. It therefore omits any trace of Laysan's human history: the encampments of bird-hunters, the packing sheds from the abandoned guano mine or, just to the south, the graves of the two murdered Japanese.

The effects of the Laysan Incident reverberated further still, for the order to establish the Hawaiian Islands Reservation set the precedent for an array of other conservation areas along the margin of the United States. During the early years of Roosevelt's presidency the bulk of bird reservations had been set up in the metropolitan US, in places within striking distance of major urban areas such as upstate New York or the Florida coastline, then a booming tourist destination

142 NARA-II MD Territories Office, Hawai'i, Bird Depredations: 'Memorandum for the Chief Clerk', 27 October 1909; Unger, *Max Schlemmer*, 64.

143 UHM WA Bryan collection, Laysan field diary 1911. It was not until twelve years later, in 1923, that a second expedition finally succeeded in eliminating rabbits from Laysan. Storrs L. Olson, 'History and Ornithological Journals of the Tanager Expedition of 1923 to the Northwest Hawaiian Islands, Johnston and Wake Islands', *Atoll Research Bulletin* 433 (1996): 29–30, 48, 55.

144 Rachel Poliquin, *The Breathless Zoo: Taxidermy and the Cultures of Longing* (University Park, PA: Pennsylvania State University Press, 2012), 104.

Figure 3.4 The Laysan Cyclorama in the University of Iowa's Museum of Natural History
© Britt Fowler, 2015

thanks to the opening of a new railway line.[145] But the Laysan Incident helped draw Roosevelt's attention to America's thinly populated colonial territory on the Pacific rim. Between issuing Executive Order 1019 and leaving office a month later, Roosevelt issued orders to establish six other bird reserves, strung along the Alaskan coastline from Sitka to the Aleutian Islands.[146] Like the Northwest Hawaiian Islands, these reservations were inacessible to all but the most well-heeled American bird-fanciers. They were, however, accessible to the bands of Japanese bird-hunters that roamed the North Pacific.

Finally, it is significant that Roosevelt's 1909 proclamation explicitly excluded the island of Midway, where the strategic imperative to protect the cable-landing station trumped the need to conserve birdlife.

[145] For the paradigmatic case of Florida's Pelican Island Reserve see Barrow, *A Passion for Birds*, 133–4.

[146] See Roosevelt Executive Orders 1037, 1039, 1040, 1041, 1044 and 1049. During the same period also established several other bird reserves in the metropolitan United States, and one in the colony of Puerto Rico. See EO 1032, 1042 and 1042. Viewable at: www.presidency.ucsb.edu/documents/app-categories/written-presidential-orders/presidential/executive-orders. Last accessed 9 February 2022.

During the Pacific War the island was, like many other coral atolls, transformed into a major naval airbase, with 'almost the entire surface of the island ... either smoothed for roads, paved for runway, excavated for bunkers, or covered by buildings and dumps'.[147] Yet, as one airman complained, Midway's albatrosses 'cared nothing for the military safety of the station. Any pilot who took off or landed on the gleaming new runways could be sure of hitting half a dozen of them, sometimes bending a prop or cracking a windshield.' In retaliation, US marines armed with two-by-fours clubbed thousands to death, 'with almost no effect on the population'. By the end of the war there were still some 163,000 of the birds nesting on the island. Albatross populations are sometimes capable of remarkable resilience.[148]

[147] W. Lance N. Tickell, *Albatrosses* (Mountfield: Pica Press, 2000), 206–7.

[148] Harvey I. Fisher, 'Populations of Birds on Midway and the Man-Made Factors Affecting Them', *Pacific Science* 3 (April 1949): 105, 109. This resilience is likely due to the species' long life cycle. Albatrosses can live for more than sixty years and spend most of their lives far out at sea. They do not return to their birthplace to nest until the age of seven or eight. Any culling operation would have to be maintained continuously over a decade or so in order successfully extirpate a breeding colony. See Bernhard Grzimek, *Grzimek's Animal Life Encyclopedia Vol. VII: Birds* (New York: van Nostrand Reinhold, 1975), 145.

4 Sand Dunes and Soldiers

From Phosphate Mining to National Defence, 1902–1939

In the first year of the twentieth century, Captain Sakuragi of the Imperial Japanese Navy landed on an uncharted island somewhere in the ocean south of Japan, inhabited only by gorillas, tigers and flocks of seabirds. He raised the national flag, named it the Island of the Rising Sun (Asahijima), and declared that it would 'forever more be the territory of the Empire of Japan'. His plan, he wrote in his diary, was to use the island as a base from which to build a futuristic 'undersea warship' that would leapfrog Western military technology and etch the reality of Japanese maritime dominance 'into the minds of European and American naval thinkers'.[1]

Captain Sakuragi was a fictional character, his exploits dreamt up by the pulp science-fiction writer Oshikawa Shunrō. But he was in good company. Three years later the novelist Erskine Childers would publish a bestselling espionage thriller in which two English yachtsmen discover a German plot to invade Britain from a secret military installation hidden among the sand dunes of the Frisian coastline.[2] In the years of broiling imperial fervour leading up to World War I, desert islands made ideal canvasses on which to project all kinds of skulduggery, derring-do and geopolitical rivalries.

Indeed, in some ways these speculative novels were eerily prophetic. By the 1930s the Japanese Navy *was* operating at least one secret submarine base, disguised as a civilian fishery, on an atoll in the South China Sea.[3] In the days leading up to the attack on Pearl Harbor, Japanese submarines took shelter in the shallow waters

1 Shunrō Oshikawa, *Kaitei Gunkan: Kaitō Bōken Kitan* (Tokyo: Bunbudō, 1900), 136.
2 Erskine Childers, *The Riddle of the Sands* (London: Smith, Elder & Co., 1903).
3 W. C. Jenk, HMS *Herald*, United Kingdom, 'Report on Suitability of Islands in China Sea for Landing Grounds' (18 April 1938): 3, reproduced in *Supplemental Written Submission of the Philippines to the Permanent Court of Arbitration, Vol. III*, Annex 374 (16 March 2015).

around the Northwest Hawaiian Islands, and some of the most decisive battles of the Pacific War were fought on or around tiny, remote islands such as Wake (1941), Midway (1942), Saipan (1944) and Iwo Jima (1945).

What factors drove the militarisation of desert islands in the twentieth century? Writers such as Oshikawa and Childers accurately predicted that innovations in submarine and aerial warfare would heighten islands' strategic value as landing strips and provisioning bases. But their prophecies were to some extent self-fulfilling ones: imperial thinking, to paraphrase Karl Rove, in part created its own reality.[4] In Chapter 2 we saw how bird island prospectors such as Tamaoki Han'emon and Mizutani Shinroku deployed the discourse of *Nan'yō* (South Seas) boosterism to enlist popular and state support for their colonisation ventures. Their successors would only elaborate on this theme as the twentieth century wore on, this time seeking an ally in the increasingly powerful Imperial Japanese Navy.

At the same time, the path from the imaginary exploits of Captain Sakuragi to the battlefields of the Pacific War was not a straightforward one. Right up until the eve of World War II, Japanese policy-makers were divided over whether or not annexing and fortifying remote islands served the national interest, especially if it came at the cost of damaging Japan's overseas interests elsewhere. For their part, bird island prospectors were often flexible enough to work within the existing international order, even devising elaborate schemes to secure support from foreign governments such as the United States or China. That they would cast in their lot with the Japanese Navy, or even the Japanese government at all, was by no means preordained. Crucially, however, the military proved able and willing to provide the resources to stage sustained habitation on bird islands in ways that mere civilian investors could not. Colonies established by plumage hunters and guano miners were hostage to commodity price fluctuations and the islands' own ecological limits, but the Navy had deep enough pockets to subsidise loss-making commercial operations almost indefinitely. Ironically, therefore, civilian bird island claimants turned to the

[4] Ron Suskin, 'Faith, Certainty and the Presidency of George W. Bush', *New York Times Magazine* (17 October 2004). On the reception of Childers' novel in Edwardian Britain see Leonard Piper, *Dangerous Waters: The Life and Death of Erskine Childers* (London: Hambledon & London, 2003), 75–6.

military for help performing the 'peaceable, continuous occupation' that would legitimise their claims under international law.

From Feathers to Phosphate: The Rise of Offshore Mining in Japan

Central to this story was the Japanese offshore phosphate mining industry. In the previous two chapters we saw how Japanese hunters fanned out across the Pacific in search of bird feathers, often decimating whole colonies of seabirds in order to sell their plumage on global commodity markets. For a number of reasons, the plumage trade peaked in the first decade of the twentieth century, and declined precipitously thereafter. Overhunting was no doubt one prime driver, but shifts in consumer tastes also had a dramatic impact. By the outbreak of World War I feathers had come to be viewed as 'fragile, dated and tawdry', and tastes shifted instead towards what would become the 1920s 'flapper' look.[5] This meant that exporting feathers overseas was suddenly not nearly so profitable.[6] The net result was that between 1909 and 1911 the value of plumage exported from Japan dropped by 40 per cent (Figure 2.4).[7] After 1915, the Japanese Ministry of Finance stopped bothering to categorise plumage as a distinct export category.[8] Feathers were still in demand domestically, but the small downy feathers that were used as stuffing for futons commanded but a fraction of the price that larger plumes had once fetched on international markets.[9]

Though plumage hunting went into permanent decline, Japanese interest in bird islands was sustained by a second wave of extractive industry: guano and phosphate mining. The emergence of this new industry had much to do with broader structural changes within Japanese agriculture. Though guano has a long history (as we saw in Chapter 2), hardly any of the deposits mined during the first global guano rush of the mid-nineteenth century ended up in Japan. Most guano mined in South America was sold to British farmers, or else to industrial-scale plantations growing nutrient-hungry cash crops such

[5] Stein, *Plumes*, 19. [6] Doughty, *Feather Fashions and Bird Preservation*, 51.
[7] 'Nichi-butsu bōeki to Yokohama', *Yokohama Bōeki Shimpō* (11–15 July 1914).
[8] Tōyō Keizai Shimpōsha, ed., *Nihon bōeki yōran* (Tokyo: Tōyō Keizai Shimpōsha, 1975), 30–1, 274.
[9] Umō Kenkyūkai, *Umō to negu*, 146.

as cotton, tobacco and sugar. From the 1890s, however, Japanese farmers also began to apply increasing amounts of commercial fertiliser to their crops, spurring demand for a range of different organic materials that had potential value as soil nutrients, ranging from herring caught in the Sea of Okhotsk to soybeans harvested from the plains of Northeast China, to bones sourced from abattoirs as far afield as India.[10] And of course bird excrement.

An array of different actors helped to develop guano and phosphate mining in Japan. Some bird-hunters, most notably Mizutani Shinroku, reinvented themselves as guano miners, hoping to eke new profits from the calcified excrement of the birds they had already slaughtered. But the Japanese state also played a key role, helping to popularise the use of commercial fertiliser among farmers and also promoting the kind of large-scale domestic production that might wean the country away from reliance on foreign imports.

One figure in particular exemplifies the entanglement of state and private interests in the development of this new industry. Tsunetō Noritaka was born in Fukuoka, but as young man won a prestigious government scholarship to the newly established Komaba Agricultural School. At Komaba, foreign experts hired from Germany and Britain instilled Justus von Liebig's principles of organic chemistry, teaching students how to evaluate soil quality according to the concentration of key macronutrients. Upon graduating in 1882, Tsunetō was seconded to work for the Soils Section of the Geological Survey Institute, part of the newly established Ministry of Agriculture and Commerce. For the next twelve years he assisted the founding head of the Soils Section, a foreign consultant from Germany called Max Fesca, in conducting a laborious survey of soil quality throughout Japan.[11]

Fesca's survey concluded that, compared to that of European nations, Japanese farmland was lacking in phosphate nutrients – a diagnosis that Tsunetō took to heart. In his 1896 tract *Phosphate in Enlightened Countries*, Tsunetō warned of the nation's increasing

[10] Higuchi, 'Japan as an Organic Empire', 140, 154.

[11] Max Fesca, *Nihon chisan ron* (Tokyo: Nōshōmusho Chisan Chōsasho, 1894); Tsunetō Toshihiko, 'Meiji no Pedorojisuto', *Pedorojisuto* 35 no. 1(1991): 68. Lectures by state-sponsored agronomists were a common feature of mid-Meiji rural life. See Ishikawa Daisuke, 'Meiji-ki jinzō hiryō tokuyaku hanbai mō no seiritsu to tenkai: Ibaragi-ken Chiba-ken chiiki no jirei', *Tochi seido shigaku* 173 (October 2001): 37.

reliance on imported fertiliser. Furthermore, he deployed statistics collated by the ministry to calculate that Japan was running a yawning nutrient deficit: while the annual harvest withdrew some 63,000 tons of phosphate from soil each year, only 41,000 tons worth of phosphate fertiliser was being produced domestically to replenish it. Tsunetō predicted that this phosphate gap would eventually come to cripple Japan's agricultural capacity: even after accounting for the small amount of imported fertiliser, the nation would still be saddled with a substantial phosphate deficit.[12] (Tsunetō did not anticipate the dramatic increase in fertiliser imports that would occur in the wake of the Russo-Japanese War.)[13]

In 1894, as part of a routine survey, Tsunetō stumbled upon a sizeable guano deposit on the southeastern coastline of Kyushu, and from this point began to investigate the possibility of mining phosphatic fertiliser to compensate for Japan's nutrient deficit. In 1897, after representing Japan at the World Geological Congress in St. Petersburg, he embarked upon a three-month worldwide fact-finding tour that took in the guano mines of Peru and Chile. The following year, Tsunetō succeeded Fesca as head of the Soils Section, and began lobbying the government to explore the possibility of exploiting Japan's domestic guano deposits on a commercial scale. His efforts bore fruit in 1901 when the ministry appointed him as director of the new Mineral Fertiliser Research Office (MFRO). But the initial search for guano proved disappointing, and after a year of surveys only one deposit on Ishikawa's remote Noto Peninsula showed any commercial promise.[14]

Then, in 1902, Tsunetō's pedological gaze began to drift offshore. The immediate trigger for this was the Marcus Island Incident, already discussed at length in Chapter 2. The news that American guano prospectors believed the deposits on Marcus had potential inspired him to despatch a geologist to the island, aboard the same naval warship that was sent to pick up the marines who had been stationed there to defend Japan's claim. The results of the assay were

12 Tsunetō Noritaka, *Rinpi chōsa hōbun dai 1-gō: Hyūga no kuni rinkō* (Tokyo: Nōshōmusho Chishitsu Chōsasho, 1896), 7–9.

13 Ōkawa Kazushi, *Nōrinsuisanshō chōki keizai tōkei: Nōringyō* vol. 9 (Tokyo: Tōyō Keizai Shinpōsha, 1965).

14 Tsunetō Noritaka, *Discovery and Exploitation of Rasa Island* (Tokyo: Rasa Kōgyō Kabushiki Gaisha, 1920), 1–2.

encouraging enough to persuade Mizutani Shinroku, who by then had largely wiped out the most valuable birdlife on Marcus, to begin mining guano on the island. Though the MFRO was nominally a research institute, Tsunetō clearly saw a role for his organisation in the direct operation of guano extraction and marketing. He quickly began arranging to market his 'samples' of Marcus guano fertiliser through the National Agricultural Association, a government-run network of organisations aimed at promoting new farming techniques, and within a year more than two thousand tons of guano had been excavated from Marcus and shipped to the mainland.[15]

Notwithstanding the apparent success of this public-private joint venture, the very next year the central government decided to abolish the MFRO completely. Tsunetō took the news bitterly. Years later he would not bother to disguise his sense of betrayal at what he viewed as the government's negligence:

> At this point, when we were making an important discovery toward phosphatic production in our southern possessions, the government may be said to have made an irretrievable mistake in doing away with the Research Office. The cold attitude of indifference shown by the government towards this development of national resources is astonishing.[16]

Soon after the MFRO was abolished, Tsunetō quit government service. In 1904 he published *On Japanese Soil*, which restated for a non-specialist audience his mentor Fesca's argument that Japanese farmland suffered from a phosphate nutrient deficit.[17] He also decided to continue his pursuit of guano phosphate, now in the private sector, and spent the next several years traveling extensively throughout Okinawa and Taiwan in search of potential deposits.[18] It was during these travels that Tsunetō became aware of an island called Rasa (J: Okidaitōjima), about four hundred kilometres southeast of Okinawa Main Island.

Rasa Island was another bird island that had been known to sailors for centuries but had hitherto been paid little heed. The name 'Rasa', meaning blank or empty, was given it by a navigator aboard a Spanish

15 Ōzawa Junji, *Guano hiryō: Minami-torishima san chōfun rinkō* (Tokyo: Zenkoku Hiryō Toritsugisho Kabushiki Gaisha, 1904), 6; Asō, *Rinkō*, 72–4.

16 Tsunetō, *Discovery and Exploitation of Rasa Island*, 3–4.

17 Tsunetō Noritaka, *Nihon dojō ron* (Tokyo: Seibidō, 1904).

18 Tsunetō Noritaka, *Minami Nihon no fugen* (Tokyo: Hakubunkan, 1910).

galleon, who had noted in the ship's log merely that it was a poor prospect for a harbour.[19] It had been annexed to Okinawa in June 1900 in response to a petition by a native of Niigata who hoped to develop it as a fishery station, but the plan fell through and by 1906 several different Japanese entrepreneurs were competing to establish a phosphate mine on the island.[20] One of these was Mizutani Shinroku, who managed to obtain a soil sample from Rasa to send to Tsunetō for analysis. Tsunetō determined that Rasa was geologically quite different to atolls such as Marcus or Laysan, whose phosphate reserves consisted of a layer of guano atop a coral base. On Rasa itself the island itself consisted of phosphate rock, resulting from a fusion of limestone rock with fossilised marine organisms.[21] For Tsunetō the results were so promising that, before even visiting Rasa himself, he set about establishing a consortium to purchase the lease and exploit the island.[22]

Mining operations began on Rasa in 1911, but had a rocky start. Shortly after the first labourers arrived, the island was hit by a storm that destroyed their accommodation and wrecked the desalinisation plant, their only source of drinking water. The men nearly died of thirst while waiting for the next boat to arrive, and then promptly demanded to return to mainland Japan. The next crew of miners did agree to stay on the island, but were soon sickened with beri-beri due to the lack of variety in their diet. Even after production got up and running initial sales were slow, as fertiliser manufacturers were reluctant to believe that the quality of Rasa's phosphate could equal that of the foreign imports that were now reaching the Japanese market.[23] Tsunetō's woes illustrated those of other Japanese producers, who struggled to compete with foreign importers. Until the outbreak of World War I domestically mined guano and phosphate rock comprised a mere 3 per cent of total sales within Japan (Figure 4.1).[24]

19 Bernhard Welsch, 'Was Marcus Island Discovered by Bernardo De La Torre in 1543?' *The Journal of Pacific History* 39 no. 1 (2004).

20 NAJ Naikaku, Kōbun ruijū 24(3) No. 18: 'Okinawa-ken Shimajiri-gun Minami Daitōjima o hanaru minami yaku 87 kairi ni aru mujintō o Okidaitōjima to mei tsuke dōgun Daitōjima no kuiki ni hennyū su' (22 September 1900).

21 Tsunetō Noritaka, *Jitsuyō hiryō no kōwa* (Tokyo: Chūgai Hiryō Yōhōsha, 1907), 142.

22 Tsunetō, *Discovery and Exploitation of Rasa Island*, 5.

23 Tsunetō, *Discovery and Exploitation of Rasa Island*, 14–15.

24 *Ibid*, 58–9.

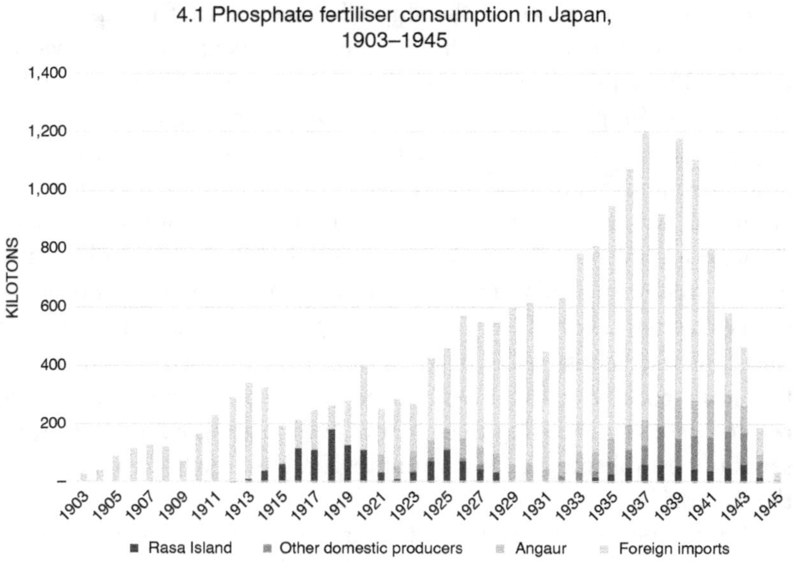

Figure 4.1 Phosphate fertiliser production in Japan, 1903–1945
Source: Kumazawa, 'Tsunetō Noritaka to Nihon no rinsan shigen.'

Warfare, Strategy and Autarky, 1914–1939

Tsunetō's company was aided, in more than one way, by the outbreak of World War I. The war's most significant boon was to create a worldwide shipping shortage that curtailed phosphate imports and sent prices spiralling. The sudden slackening in foreign competition represented an opportunity for domestic manufacturers: existing firms began ramping up their capacity and newer firms also entered the market, until by the end of the war two thirds of Japanese phosphate purchases were being produced locally.[25] Market conditions were buoyant enough for Tsunetō to expand his company's operations on Rasa throughout the war years, until by 1918 there were more than 2,000 workers on the island mining over 160,000 tons of phosphate per year (Figure 4.2).[26]

The war also provided the Rasa Company with a powerful new ally in the form of an increasingly well-financed and assertive Imperial Japanese

[25] Kumazawa Kikuo, 'Tsunetō Noritaka to Nihon no rinsan shigen', *Hiryō Kagaku* 6 (1983): 58–9.

[26] *Ibid*, 37–8; Nihon Kasei Hiryō Kyōkai, *Rinsan hiryō kōgyō no ayumi* (Tokyo: Nihon Kasei Hiryō Kyōkai, 1972), 42–3.

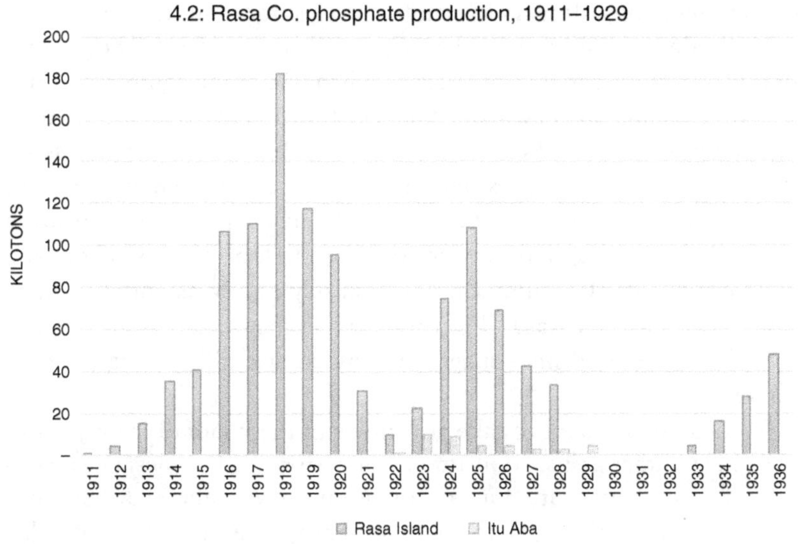

Figure 4.2 Rasa Co. phosphate production, 1911–1929
Source: *Rasa Kōgyō 80-nen shi*

Navy willing to mobilise public opinion in support of *Nan'yō* expansionism. Shortly after the outbreak of World War I in Europe, Japan seized the opportunity to 'liberate' Germany's colonies in Micronesia. After the war the invasion was retroactively legitimised by the newly established League of Nations, which charged Japan with preparing Micronesia for independence at some unspecified date in the future.[27] In fact, the Navy set about subsidising large-scale Japanese migration to the islands. By 1940 some 77,000 Japanese had migrated to the South Seas Mandate, on some islands outnumbering Indigenous Micronesians by a ratio of nine to one.[28] It also began bankrolling a new generation of boosters to whip up public enthusiasm for *Nan'yō* colonialism, hoping to secure larger budget allocations from the central government.[29]

Tsunetō and his new company took full advantage of this renewed wave of *Nan'yō* interest. In 1913 he had unsuccessfully petitioned the

[27] For a discussion of Japanese rule in Micronesia, see Peattie, *Nan'yō*; Tsujihara, Makihiko, Imaizumi, Yumiko, *Nan'yōchō Kohō* (Tokyo: Yumani Shobō, 2009); Ti Ngo, 'Mapping Economic Development: The South Seas Government and Sugar Production in Japan's South Pacific Mandate, 1919–1941', *Cross-Currents: East Asian History and Culture Review* 2 (March 2012).

[28] Peattie, *Nan'yō*, 160. [29] Schencking, *Making Waves*, 213.

Communications Ministry to install a wireless radio station on Rasa, so as to obtain more accurate updates on weather conditions. The urgency of this was highlighted in June 1914, when one of the company's ships was wrecked in a storm with all hands lost. After Japan declared war on Germany in August, Tsunetō repeated his request for a wireless on Rasa, emphasising 'the strategical importance of such a means of communication' during wartime. Crucially, in his renewed application Tsunetō also managed to convince the Navy to lobby on his behalf, and in December the Communications Ministry finally assented to open a wireless office on Rasa.[30] In this way, Tsunetō successfully argued that the interests of his young company coincided with the national interest.

Tsunetō also sought to take advantage of the windfall from Japan's wartime military gains. Soon after the invasion of Micronesia he placed a bid for the phosphate mine on the island of Angaur, that had been confiscated from the German South Sea Phosphate Co.[31] The bid failed but Tsunetō did convince the Navy to second officers to lead three successive expeditions to survey guano islands in the South China Seas.[32] The expeditions were conducted between 1918 and 1921, nominally under the private auspices of the Rasa Company, so that they could not be construed as official state ventures.[33] But they were a clear gesture of military support for Tsunetō's efforts to claim new *Nan'yō* territory for the Japanese Empire.

World War I left another legacy. In the interwar period bird islands played an increasingly important role in the strategic thinking of the

30 Tsunetō, *Discovery and Exploitation of Rasa Island*, 26.

31 NAJ Kōbun zassan Taishō 5-nen dai 24-hen dai-35-kan: 'Rasa-tō Rinkō Kabushiki Gaisha shachō Tsunetō Noritaka kaishin Nan'yō Angauru-jima jigyō keiei kyoka seigan no ken' (19 June 1916). Instead, the licence to mine Angaur phosphate was granted to Nishizawa Yoshiji, the same man who had inflamed Chinese public opinion by staking a claim for Pratas Island (see pages 138–9). Nishizawa operated the Angaur concession for more than a year before the Navy revoked his licence, worried that his maltreatment of Indigenous workers would undermine Japan's claim to be liberating Micronesians from German oppression. (Nishizawa was rumoured to have been physically abusing his workers and paying them with currency he printed himself.) See Peattie, *Nan'yō,* 67.

32 MOFA A.4.1.0.2–1–1 Vol. I: No. 4 'Kinkai Dai-2 kuiki Minami Shina Kai chū Shin Nan Guntō kōkō rinkō tsumitori-sen naikō fune toriatsukai ni tsuki seigan' (5 June 1923).

33 Rasa Kōgyō Kabushiki Gaisha Shashi Henshūshitsu, *Rasa Kōgyō 80-nen shi* (Tokyo: Rasa Kōgyō Kabushiki Gaisha, 1993), 45.

Japanese Navy. Like many others, Japanese strategic thinkers chose to attribute Germany's defeat in World War I to the success of blockade warfare. According to this theory, the *Wehrmacht* was not defeated on the front lines, but was instead slowly starved of resources due to the Royal Navy's maritime dominance in the North Sea. Recent historiography has begun to cast doubt on this interpretation, but for both branches of the Japanese military it had an obvious appeal – if only because it could be used to justify larger budget appropriations.[34] For the Imperial Japanese Army, the blockade theory proved that Japan must gain and maintain control over the rich coal mines, abundant farmland and rapidly industrialising economy of Northeast China. And the Navy used the same premise to draw the opposite conclusion. Japan, successive naval ministers argued, should prioritise a 'southern advance' to secure the oil wells and rubber plantations of Southeast Asia, and then establish naval supremacy in the West Pacific in order to defend the sea lanes connecting its new conquests with the home islands.[35]

Keeping the Peace in the South China Sea

But while army and naval officers nurtured dreams of expansive military adventurism, Japan's Ministry of Foreign Affairs preferred to pursue a policy of interimperial cooperation. We have already seen how the Foreign Ministry generally pursued a policy of accommodation with the United States in order to advance Japan's non-territorial interests. Although it successfully asserted Japan's sovereignty over Marcus Island, it recognised the United States' claim to other North Pacific bird islands lying as far west as the Micronesian atoll of Wake. And it made sure to cultivate relations with other Pacific imperial powers. It entered into a formal naval alliance with Great Britain in 1903, and in 1905 it invited President Theodore Roosevelt to broker its peace with Russia. It ensured that Japan sided with both powers in the war against Germany, and after the war it signed a trilateral treaty in

[34] Michael A. Barnhart, *Japan Prepares for Total War: The Search for Economic Security, 1919–1941* (Ithaca: Cornell University Press, 2001) 23; Schencking, *Making Waves*, Ch. 8; Nicholas Mulder, 'The Economic Weapon: Interwar Internationalism and the Rise of Sanctions, 1914–1945', PhD diss., Columbia University, 2019).

[35] Barnhart, *Japan Prepares for Total War*. 28, 168.

which all three states agreed to limit fleet sizes to a fixed ratio of battleship tonnage.

The ministry also trod carefully in relation to China, where Japanese foreign policy sought to reconcile the tension between acquiring and preserving territorial colonies and promoting its citizens' commercial interests more broadly. By 1900 Japan had become a fully paid up member of the imperial club, having not only seized the island of Taiwan from China but also participated in an eight-nation alliance to suppress an anti-colonial uprising in Beijing.[36] As such, the country quickly became a target of Chinese nationalist anger. The most famous example of this came during World War I, when the Japanese government presented China with a list of twenty-one demands aimed at transforming the country into a de facto client state. This attempt at diplomatic strong-arming ended up being counter-productive: it triggered a nationwide boycott of Japanese goods and played a significant part in the outbreak of China's May Fourth Movement.[37] For much of the 1920s Japanese diplomats struggled to balance what they saw as their national security needs, which in practice often involved flagrant violations of Chinese sovereignty, with the desire to avoid adversely impacting their citizens' commercial interests there.[38]

A similar tension, between advancing Japanese territorial claims and maintaining good relations with China, also shaped policy toward bird islands. In 1907 a prospector by the name of Nishizawa Yoshiji submitted an inquiry to the Japanese consulate in Hong Kong about the Pratas Islands (Ch: Dongsha Qundao), one of several groups of cays and atolls scattered throughout the shallow basin of the South China Sea. In geographical terms the islands were closest to China's Guangdong Province, but consular officials informed Nishizawa that the islands were to the best of their knowledge neither British nor Chinese territory. Nishizawa then petitioned the Japanese Home Ministry to annex the islands to Taiwan, but the Foreign Ministry

36 Rana Mitter, *A Bitter Revolution: China's Struggle with the Modern World* (Oxford: Oxford University Press, 2005), 33–4.

37 Mitter, *A Bitter Revolution*, 36.

38 Ian Nish, 'An Overview of Relations between China and Japan, 1895–1945', *The China Quarterly* 124 (December 1990): 612; Banno Junji, 'Japanese Industrialists and Merchants and the Anti-Japanese Boycotts in China, 1919-1928', in *The Japanese Informal Empire in China, 1895–1937*, ed. Peter Duus, Ramon H. Myers and Mark R. Peattie (Princeton: Princeton University Press, 1989), 320.

expressed concern that such a move might encounter British opposition. Instead, it advised Nishizawa to go ahead and begin mining guano on the islands, in the hope that by occupying them he might strengthen the case for formal annexation at some point in the future.

Two years later, in 1909, news that Nishizawa was mining on the Pratas Islands hit the headlines in China, sparking threats of a boycott of Japanese goods. The protests appear to have caught Japanese diplomats by surprise. Up until that point they had only worried about offending Britain, not China, after all, and consular officials believed that in terms of international law Chinese claims to the islands were weak. By this point Nishizawa had already sunk substantial capital into his venture – including piers, telephone cables, minecart tracks and dormitories for labourers. He had also developed a profitable side-business processing turban-shells, seaweed, plumage and the shells of the hawksbill turtles that crawled ashore to lay their eggs on the island.[39] Nishizawa's claim to have established continuous occupation of Pratas was therefore at least as plausible as Mizutani's claim to Marcus. Still, rather than risk aggravating relations with China further the Foreign Ministry decided to seek a negotiated settlement, and persuaded Nishizawa to withdraw from the islands in return for an indemnity of 130,000 silver dollars paid by the Guangdong government.[40]

We can see another example of this relatively conciliatory stance in the decade after World War I. In 1920 another Japanese national named Hirata Sueharu began mining phosphate on the Paracel Islands (Ch: Xisha Qundao, V: Quần đảo Hoàng Sa), a smattering of South China Sea reefs and cays, roughly equidistant from China and Vietnam.[41] Hirata was originally from Osaka but had established himself as a medium-sized player in the merchant community of the southern Taiwanese port city of Takao (Ch: Kaohsiung, Rukai: Takau). A report on Hirata compiled by the Taiwanese colonial government described him as 'character with a certain kind of charm' who was involved in a wide range of South China Sea industries including

39 NIDS Kaigunshō M41-107–908: Gunmu-kyoku, 'Puratas-tō shisatsu hōkoku' (1908).

40 Bill Hayton, *The South China Sea: The Struggle for Power in Asia* (New Haven: Yale University Press, 2014), 51.

41 Xu Longsheng, 'Zhong-Ri liang guo weirao Xisha Qundao lin kuang kaifa de hezuo, jingzheng yu jiufen (1917–1930)', *Shi Lin* 5 (2017): 117–18.

guano mining, fishing and the gathering of other valuable marine-products.[42] Initially Hirata hoped to follow the precedent established by Tamaoki Han'emon and Mizutani Shinroku: establish an economic presence on the island and then lobby the Japanese government to annex it. But when the Foreign Ministry got wind of the scheme in January 1921 it immediately ordered him to cease operation on the Paracels lest it trigger another diplomatic spat with China.

Instead, the ministry encouraged Hirata to secure permission to mine from the Chinese government instead. In an elaborate scheme concocted with help from the Japanese consul in Canton and the Taiwanese colonial government, Hirata partnered with a Chinese merchant by the name of Liang Guozhi, who applied to the Guangdong provincial government at Canton for permission to mine guano on the Paracels. It was not, at that point, at all obvious that China considered the Paracels its own. In 1895, for instance, the Guangdong government had shrugged off a British insurance company's attempt to claim compensation for a shipwreck that had been looted by Hainanese fishermen, 'on the grounds that, for them, the Paracels were abandoned islands that belonged no more to China than to Annam, and that 'no special authority was responsible for policing them'.[43] But the prospect of securing licensing fee revenue induced the government to reclaim the islands once more.

The mining venture quickly became mired in Guangdong factional politics, however. A rival Chinese merchant called He Ruinian, who boasted ties to no less exalted a personage than Sun Yat-sen himself, began contesting Liang's claim to the Paracels, and succeeded in securing a provisional mining permit.[44] Hirata responded by essentially cutting He into the deal as well, and by December 1921 he had managed to resume mining on Pratas. Only a few months later, however, student groups with ties to the Chinese nationalist party published a salvo of newspaper articles accusing Liang and He of selling out the nation to Japanese imperialism. In the process it emerged that Liang, although ethnically Chinese, had in fact been born in the British colony of Hong Kong. The scandal triggered an investigation by the Canton

[42] MOFA A.4.1.0.2–1–1 Vol. I: No. 8; Hirata Sueharu, 'Shin Nan Guntō no konjaku' in *Taiwan Jihō* (20 May 1939).

[43] Monique Chemillier-Gendreau, *Sovereignty Over the Paracel and Spratley Islands* (Leiden: Brill Nijhoff, 2000), 36–7.

[44] Xu, 'Zhong-Ri liang guo weirao Xisha Qundao lin kuang kaifa', 118–19.

Bureau of Public Security, which eventually led to the company's licence being suspended.

Hirata and his business partners responded to this setback creatively. They got the Guangdong government to hastily affirm that Liang was, in fact, a citizen of the Republic of China. And with the help of the Taiwanese colonial government and the Japanese consul in Guangdong they hammered out a deal whereby the company promised to hire Hainanese islanders instead of Okinawans, thereby ensuring that more revenue from the mining operation ended up in Chinese pockets. (Evidence suggests that Hirata did not in fact honour this promise, and continued to hire Japanese and Taiwanese workers.)[45] These manoeuvres served to mollify Chinese nationalist sentiment enough to keep the mine running, with only occasional interruption, until the end of the decade.[46]

The Pratas guano mining venture is interesting for two reasons. First, it shows that Japanese prospectors in the South China Sea were flexible enough to work with foreign business partners, and indeed foreign governments, in order to secure the rights to bird islands. In some cases, they did so in order to compete with their own compatriots. In one remarkable incident in 1927, a Hong Kong-based Japanese merchant by the name of Obata Masakazu secured the rights to Pratas from the Guangdong provincial government and then made sail for the island accompanied by thirty armed Chinese mercenaries. Declaring that the mine and all its equipment had been confiscated by the Guangdong government, Obata ordered the Japanese mining crew to leave and set about mining the island with his own crew of Chinese labourers.

Second, and just as interesting, was the generally conciliatory stance struck by Japanese diplomats, who preferred to work with Guangdong government officials (thereby recognising Chinese sovereignty over the islands) rather than jeopardise Sino-Japanese relations. In the wake of Obata's raid the Foreign Ministry refused calls to send a retaliatory gunboat to Pratas, even though at the time the main investor in the mine was a Japanese Diet member.[47] The Diet member lost his stake and He Ruinian eventually regained control of the mine, this time with financing from yet another Japanese commercial backer.

[45] *Ibid*, 120–1, 123. [46] *Ibid*, 125. [47] *Ibid*, 123–4.

Islands in the Dark Valley

This relatively dovish stance towards South China Sea island disputes did not survive into the 1930s, however. In 1931 the Japanese Army decided, without consulting the Foreign Office, to invade Northeast China and set up a Japanese-run puppet state. This unsurprisingly plunged Sino-Japanese relations into a deep-freeze from which they did recover until after World War II. In the years that followed the military came to play an increasingly central role in Japanese politics: between 1931 and the end of the World War II the army and navy each supplied four of Japan's fourteen prime ministers, and after 1936 it became customary for ministers of the Army and Navy to be appointed from the ranks of serving officers.[48] The 1930s therefore saw a struggle for control over Japanese foreign policy between the old guard of Foreign Ministry diplomats and a newer, more hawkish breed of military man.[49]

Alongside all this geopolitical jockeying and interdepartmental squabbling, the interwar period also saw a more gradual shift in strategic thinking about the value of remote islands. World War I had stimulated combatant states to research two other technologies that relied on islands as refuelling depots: namely, submarines and aeroplanes. Ironically the post–World War I peace settlement helped to accelerate research into these emerging technologies, because both were exempt from the trilateral arms limitation treaties signed by Japan, Britain and the United States. Beginning in 1924 the Japanese Navy therefore poured research funds into a variety of new submarine designs, including single-person 'midget' submarines and vessels capable of launching reconnaissance aircraft. After the 1930 London naval treaty capped submarine tonnage as well, it began ramping up aircraft production instead, partly in an effort to keep pace with the expansion of US naval air power (Figure 4.3). By the eve of World War II it had developed a land-based bomber with a range of 2,300 miles – roughly as far as the feeding range of a Steller's albatross.[50]

[48] Gordon Berger, *Parties out of Power in Japan, 1931–1941* (Princeton: Princeton University Press, 1977).

[49] For a discussion of the degree to which 1931 marked a turning point in Japanese foreign policy see Erik Esselstrom, *Crossing Empire's Edge: Foreign Ministry Police and Japanese Expansion in Northeast Asia* (Honolulu: University of Hawai'i Press, 2009) 3–4, 156–7.

[50] Yōichi Hirama, 'Japanese Naval Preparations for World War II', *Naval War College Review* 44 no. 2 (1991) 66, 68–70. Steller's albatrosses have been known to

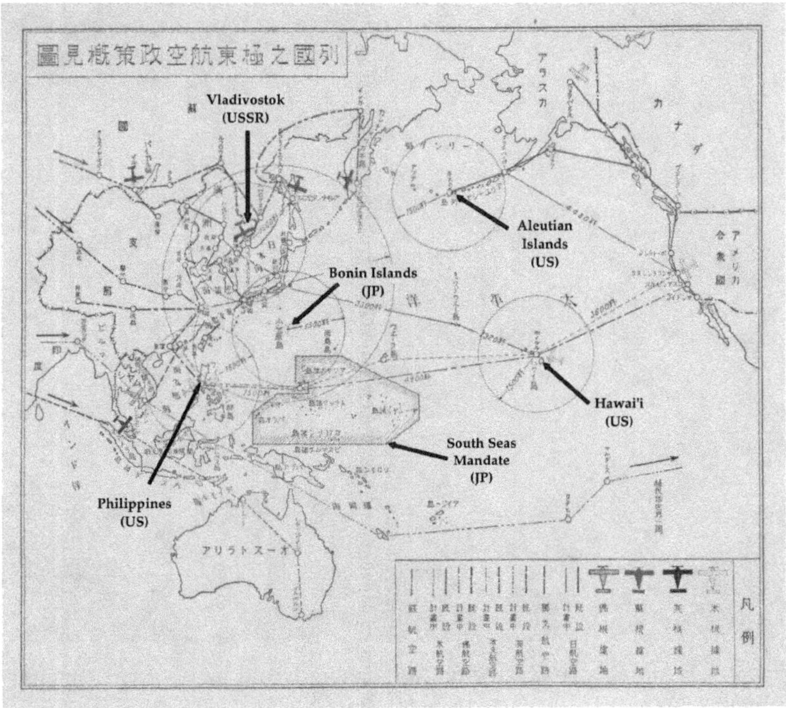

Figure 4.3 'Far Eastern Aviation Policy of the Great Powers' (1936)
Source: Uyama Kumatarō, *Kokubō-ron* (1936)

For both planes and submarines, with their relatively limited fuel capacity, remote islands made ideal refuelling bases. Atolls were particularly valuable: their friable soil and flat coral surfaces could be coated with cement to form landing strips (Figure 4.4), and the shallow waters surrounding them could conceal submarines from detection by sonar.[51] As a result, militaries in many countries began to pay increased attention to the potential strategic value of remote islands as bases.[52] As early as 1930 the Royal Air Force was appraising the atolls in the South China Sea as 'possible, though by no means ideal, refuelling points for

feed as far as 1,988 miles from their breeding grounds: https://www.fws.gov/species/short-tailed-albatross-phoebastria-albatrus. Last accessed 20 May 2020.

51 Peattie, *Nan'yō*, 231.

52 UKNA T161/622: J. S. Barnes to Foreign Office, 6 May 1930; 'Memorandum' (n.d.); J. S. Ross to Foreign Office (5 May 1931).

Figure 4.4 Airfield constructed on Middleburg Island by US Army engineers (August 1944)
Source: National Archives and Records Administration digital collection

aircraft'.[53] From 1931 Japan began building airstrips on the Bonin Islands (including on Marcus Island) and also across Micronesia – in defiance of League of Nations rules against militarising mandate colonies). As war in East Asia loomed other countries followed Japan's lead.[54] By 1936 Britain and China began discussing the possibility of building a military airstrip on the Paracel Islands in order to aid the defence of Hong Kong and Singapore, and the US likewise began to fortify uninhabited Pacific islands such as Baker, Jarvis and Midway.[55]

53 UKNA REFUKNA T161/622: J. S. Barnes to Foreign Office, 6 May 1930; 'Memorandum' (n.d.); J. S. Ross to Foreign Office, 5 May 1931).
54 Ishihara Shun, *Iōtō: kokusaku ni honrō sareta 130 nen* (Tokyo: Chūkō Shinsho, 2019): 88; *Peattie*, 231–3; Eldridge, *Iwo Jima and the Bonin Islands in US–Japan Relations*, 104.
55 MOFA A.4.1.0.2–1–1 Vol. 2: Japanese Consul in Saigon, 'Paraseru Guntō ni kan suru Ei_Shi kōshō to Butsu oyobi Inshi no sochi' (31 October 1936); Immerwahr, *How to Hide an Empire*, 341.

Sovereignty and Strategy on the Spratly Islands, 1918–1939

For an example of how all these processes helped shape the history of bird islands, we can return to the Spratly Islands. As we saw in Chapter 2, in the 1870s the Spratlys had briefly been the site of a calamitously unsuccessful Anglo-American guano mining venture. But during the interwar period the islands became the object of a protracted diplomatic dispute between the governments of France and Japan. The dispute was not only inter-governmental, however. It also pitted two separate branches of the Japanese government against one other, and shows how distinct Japanese commercial interests sought to mobilise support from different state institutions in order to assert control over bird islands, against a background of shifting strategic understandings of the value of remote islands.

The first recorded instance of Japanese interest in the Spratly Islands dates to 1918, when Tsunetō Noritaka's Rasa Company notified the Naval, Colonial and Foreign Ministries that it had 'discovered' an archipelago of desert islands that boasted particularly promising phosphate reserves. 'Discover' was a pure legal fiction, for the islands had been well known to navigators for hundreds of years. The company's own survey showed extensive evidence of inhabitation by Chinese fishermen, who had dug several wells on one of the islands and even constructed a small shrine. Regardless, the Rasa Company christened the islands the 'New Southern Archipelago' and petitioned for them to be recognised as Japanese territory.[56]

The Foreign Ministry responded to this request warily, suggesting that the company first establish its presence on the islands as a way of strengthening the case for annexation. There are various ways of interpreting this stance. On the one hand it drew on the theories of international law discussed in Chapter 2 that predicated sovereignty over territory on continuous occupation it. But it was also a way to avoid antagonising the Chinese government unnecessarily. Following this advice, in 1921 the Rasa Company set about establishing a token mining operation on the island of Itu Aba (Ch: Taiping-dao, J: Nagashima). Itu Aba's output was but a fraction compared to the volume of phosphate extracted from the company's main mine on Rasa Island (Figure 4.2), but its purpose was not so much to earn

56 Rasa Kōgyō Kabushiki Gaisha Shashi Henshūshitsu, *Rasa Kōgyō 80-nen shi*, 45–9.

immediate revenue as to demonstrate continuous occupation of the islands – to foreign governments, but also, it should be said, to rival Japanese mining outfits as well.[57]

The Rasa Company's decision to stake a claim to the Spratlys was in large part a product of the wartime boom in domestic Japanese phosphate production. But the return of peacetime conditions to the phosphate industry showed the company's expansionist zeal to have been misplaced. During the war the company had been insulated from foreign competition; but the post-war recession brought about an abrupt realisation that the industry as a whole was now characterised by chronic overproduction (Figure 4.1).[58] The Rasa Company suddenly found its product competing with rock sourced from as far afield as Egypt and Florida, but it had borrowed heavily to fund its wartime expansion, including the Itu Aba mine, and now found itself crippled with debt.[59] As the price of phosphate remained stagnant throughout the 1920s, the company's fortunes continued to wane. It gradually reduced the scale of its operations on both Rasa Island and Itu Aba, until finally in 1928, as part of a comprehensive debt restructuring package, it switched the focus of its business from mining raw phosphate to manufacturing branded fertiliser using phosphate purchased from the South Seas Mandate government's mine on Angaur. The mining facilities on both Rasa Island and Itu Aba were mothballed until such a time as they might prove profitable once more.[60]

At around the same time the Rasa Company was winding down its operations in the South China Sea, other prospectors from outside Japan were beginning to take an interest in the region. Beginning in 1925 the Institute of the Oceanographic Service of Indochina despatched a series of cruises to survey both the Paracel and Spratly Islands. The institute's researchers were primarily interested in studying coral formation, but to secure further government funding they drummed up the islands' economic importance, including by inviting

57 *Ibid*, 45–9.
58 *Ibid*, 49. By 1921 Japan's fourteen phosphate manufacturers between them had an annual capacity of 32 million tons, double the amount purchased yearly by Japanese farmers.
59 'Rinkōseki no jukyū akka', *Keizai Zasshi Daiyamondo* (1 March 1927): 19; 'Rasa-tō Rinkō no kinkyō', *Keizai Zasshi Daiyamondo* (11 July 1927): 46–7.
60 'Rasa-tō Rinkō Kaisha', *Keizai Zasshi Daiyamondo* (1 May 1929): 83.

the New Phosphates Company of Tonkin to send a chemist to collect guano samples.[61]

Shortly afterwards the company began lobbying for a licence to mine on the Spratlys, and over the next few years officials representing various branches of the French state began advancing territorial claims to the islands. In 1929 the Governor of Cochin China declared the Spratly Islands were henceforth to be administratively attached to Baria Province. In April 1933 French naval forces sailed to the Spratlys, unfurled the French flag and deposited on each island a title of possession sealed inside a glass bottle, itself incased inside a boundary marker cemented to the island floor. Finally, in July of that year, the French government gazette formally proclaimed the annexation of the Spratly Islands by the French Navy.[62]

News of the French annexation roused Rasa Company executives to action. By 1933 the Japanese economy was rebounding from the depths of the Great Depression, and phosphate prices were finally starting to rally after a decade in the doldrums.[63] The company was anxious not to let its claim to Itu Aba slip just at the point when its efforts might be about to pay off. Frustrated at the Foreign Ministry's reluctance to press the issue, the Rasa Company set about rallying public sentiment to bolster its case. In August 1933 Asanuma Fujinosuke, a former Rasa executive, arranged for two *Osaka Mainichi Shimbun* journalists to travel secretly to Itu Aba and reconnoitre the situation on the spot.[64] These reporters returned from their trip waxing lyrical at the Spratlys' geopolitical significance – 'the Balkans of the South China Sea' – and their boundless economic value: 'Small though [Itu Aba] may be … there are still considerable reserves of guano and phosphate to be mined … [and] as a nation of fishermen how can Japan bear to forsake the New Southern Archipelago?'[65] The *Osaka Mainichi* report prompted further coverage

61 Gerard Sasges 'Absent Maps, Marine Science, and the Reimagination of the South China Sea, 1922–1939', *The Journal of Asian Studies* 75 no. 1 (February 2016): 169.

62 Chemillier-Gendreau, *Sovereignty Over the Paracel and Spratley Islands*, 37–8, 64.

63 MOFA A. 4.1.0.2–1–1 Vol. I: No. 5; 'Shin Nan Guntō ni Nichi-Butsu no ryōdoken mondai okoru', *Tokyo Nichinichi Shimbun* (30 July 1933).

64 MOFA A.4.1.0.2–1–1 Vol. I: No. 6.

65 Matsuo Kunizō & Miyoshi Takeji, 'Miyo, itaru tokoro "Nihon" no ashiato wa yakudō', *Osaka Mainichi Shimbun* (6 September 1933).

extolling the value of the Spratly Islands to Japan, and the Foreign Ministry duly swung round to defend the Rasa Company's claim, now protesting that the French annexation disregarded the islands' prior development by Japanese nationals.[66]

Yet even as the Foreign Ministry was shifting to back the Rasa Company, the Naval Ministry began to worry whether a mere 'profit-seeking' company could be trusted to maintain a permanent settlement in the New Southern Archipelago, as in the face of another economic recession it might simply decide to up sticks and leave once more.[67] Instead, it hit upon another scheme. Together with the Government General of Taiwan it bankrolled a fisheries operation on Itu Aba through a civilian proxy.[68] The scheme offered several advantages. Fisheries promotion in the Spratlys was a relatively straightforward business. It was not particularly capital-intensive, and it provided the perfect cover for installing infrastructure necessary for 'national defence' – a wireless station, meteorology equipment, 'protection against pirates' and, most importantly, oil storage tanks.[69] Another memo put it more baldly: a fisheries station on Itu Aba would also allow the island to serve as a submarine refuelling station.[70]

As a frontman for its operation the Navy turned to none other than Hirata Sueharu. We last encountered Hirata in 1921, when he had covertly secured a license to mine guano on the Paracel Islands from the Guangdong provincial government of China. The failure of this scheme drove Hirata into bankruptcy, but over the course of the 1920s he gradually resuscitated his fortunes while cultivating relationships with

66 Yokota Kisaburō, 'Mujintō sensen-ron', *Chūō Kōron* (September 1933): 84–93; MOFA A.4.1.0.2–1–1 Vol. I: No. 7.

67 MOFA A.4.1.0.2–1–1 Vol. I: No. 12.

68 MOFA A.4.1.0.2–1–1 Vol. I: No. 10. Shirane suggests that the Taiwanese colonial government had its own ambitions in the South China Sea, for it hoped that a Spratly annexation might lead to the creation of a new, Taiwan-led bloc that would include its arch-rival the South Seas Mandate as well. See Seiji Shirane, 'Mediated Empire: Colonial Taiwan in Japan's Imperial Expansion in South China and Southeast Asia, 1895–1945' (PhD diss. Princeton University, 2014), 184.

69 'French Flag over the Unoccupied Islets', *The Illustration* (15 July 1933): 382; C(T)SIL Amicus Curiae Submission, Exhibit 33; P. J. Philip, 'Barren Coral Isles in Pacific Cause International Dispute', *New York Times* (27 August 1933): 1; SWSP, Vol. IV, Annex 401; 'Shin Nan Guntō no kankatsu kettei kyō kōkoku', *Osaka Asahi Shimbun* (18 April 1939); MOFA A.4.1.0.2–1–1 Vol. I: No. 6.

70 MOFA A.4.1.0.2–1–1 Vol. I: No. 12.

Japanese naval officers based in Taiwan.[71] In 1933 he made the pages of the *New York Times* by agitating for Japan to annex the Paracels, which he immodestly suggested be renamed the 'Hirata Archipelago'. Much like the Rasa-backed newspaper reporters, Hirata did his utmost to emphasise the value of the islands to Japan, emphasising their proximity to Hong Kong, Macao and the Chinese naval port of Hainan.[72] In this sense Hirata was more than just a frontman for the Japanese Navy. He also provided valuable intelligence, gleaned from his various escapades around the South China Sea, and used his skills as a booster to whip up public enthusiasm for the Southern Advance.

The Foreign Ministry was unenthusiastic about this new joint venture, which it (correctly) perceived as yet another attempt by the military to seize the initiative in setting Japan's foreign policy. It noted with chagrin that Hirata's company, although primarily a fisheries concern, was also claiming de facto rights to mine phosphate on Itu Aba, overriding any claim that the Rasa Company had built up during its years working the island. In a top-secret internal memo dated September 1937 a Foreign Ministry official lamented that, unlike the Rasa Company, 'Hirata has no previous connection at all with the islands; the Navy's support is regrettable.'[73] But there was only so much it could do. The ministry did manage to have the Ministry of Finance veto the Taiwan Government General's planned ¥100,000 subsidy to Hirata's company, but was unable to stop the Navy from providing double that amount.[74]

Kaiyō Kōgyō's fishery station began operation on Itu Aba in 1935. According to the accounts circulated to the Foreign Ministry in its first year, the operation did generate a substantial revenue in sales of marine products but recouped only 75 per cent of its investment.[75] It seems unlikely the fishery ever turned a profit. But then again the Rasa Company's earlier guano mine had not done so, either, and like that earlier operation, the fishery station's immediate purpose was not to make a profit but to lay claim to the island as both property and territory. A secret memo forwarded to the Naval Ministry by Hirata

71 MOFA A.4.1.0.2–1–1 Vol. I: No. 8.
72 'New Island Claim Urged by Japanese: Exploitation Group Calls on Tokyo to Annex Group in North of China Sea', *New York Times* (10 September 1933).
73 MOFA A.4.1.0.2–1–1 Vol. I: No. 24.
74 MOFA A.4.1.0.2–1–1 Vol. I: No. 10; No. 11.
75 MOFA A.4.1.0.2–1–1 Vol. I: No. 8.

refers to Itu Aba as Japan's 'front-line base' in the South China Sea, and reveal that the eleven 'fishermen' on the island reported to a captain in the navy's No. 5 Torpedo Squadron.[76]

In 1937 the Japanese army launched a full-scale invasion of China; and by the summer of 1938 it was becoming increasingly likely that Japanese troops would as far south as Guangdong, if not further. In August that year a French naval vessel disembarked personnel on Itu Aba, no doubt hoping to reassert French control over the island in case it proved strategically useful in the event of war with Japan. The French proceeded to construct a rival wireless station, meteorological station and lighthouse.[77] As the two forces faced each other on the tiny island, the Japanese Naval Ministry responded to a French complaint about Japanese 'trespassers' by stating that the fishery on Itu Aba operated there with the protection of the Japanese government, and that any interference with them would necessarily demand a response.[78]

The fisheries base on Itu Aba therefore gave the Navy a pretext to wrest control of foreign policy away from civilian diplomats and set a more assertive stance vis-à-vis France. Finally, after the invasion of Hainan in February 1939, the Naval Ministry was able to convince the Foreign Ministry to declare a formal annexation of the Spratlys, which were henceforth to be incorporated into Takao, Taiwan. The rights to operate on Itu Aba were retroactively granted to Hirata's company, with the Rasa Company's complaints falling on deaf ears. There was a certain irony to this, as the company itself had played up the strategic value of the Spratly group, hoping that it would serve its own interests. Now, however, strategic objectives had decisively trumped commercial logic, and the company found itself cut out of the loop altogether.

In the event, Japan's declaration of sovereignty over the Spratly Islands met with a salvo of international condemnation. As no foreign government recognised the annexation, its only effect was to generate a tide of jingoistic media coverage within Japan itself. Newspaper articles planted by Hirata Sueharu and Tsunetō Noritaka lauded the crucial economic and strategic value of this latest southernmost addition to the Empire of Japan.[79] As the *Osaka Mainichi Shimbun*

[76] NIDS 1–391 *Sentō shōhō senji nisshi* 146: Dai-5 Suirai Sentai (December 1941).
[77] Chemillier-Gendreau, *Sovereignty Over the Paracel and Spratley Islands*, 39.
[78] MOFA A.4.1.0.2–1–1 Vol. I: No. 12.
[79] Hirata Sueharu, 'Shin Nan Guntō no konjaku', *Taiwan Jihō* (20 May 1939).

declared breathlessly, the acquisition of the Spratly Islands 'extends our national defence perimeter by one thousand miles.'[80]

It is unclear whether the Spratly Islands in fact served any useful military purpose in the war that followed. The No. 5 Torpedo Squadron did go on to participate in the Japanese invasion of the Philippines, but it did so not from the Spratlys but from Taiwan. Surviving evidence in the National Defence Archives contains no mention of any military actions launched from Japan's 'advance base' on Itu Aba.[81] During the war another Navy-financed national defence company did, briefly, reopen the island's guano mine. But its attention soon shifted to larger, more promising phosphate mines that had been seized from the British on Nauru and Christmas Island.[82] By the end of the war, there appears to have no longer been any Japanese presence on Itu Aba.

This was the case elsewhere as well. As we have seen, in the eighty-odd years leading up to World War II Japanese colonists established settlements on a broad arc of islands stretching from the North Pacific to the South China Sea. At first boosters such as Tamaoki Han'emon, Mizutani Shinroku and Tsunetō Noritaka had appealed to vague dreams of imperial expansion in order to secure state support for their own commercial ventures. Over time the islands' strategic importance had gradually come to outweigh their economic value – to the point at which civilian activities on the island were increasingly either abandoned or propped up through covert military subsidy. Finally, the devastation wreaked by Japan's disastrous war spelled the end for those few colonies that remained, so that by 1945 virtually all the desert island settlements that had been founded over the previous century had been abandoned. Japanese efforts to colonise desert islands had, in a certain sense, come full circle.

80 'Wagakuni bōsen sen-ri enchō', *Osaka Asahi Shimbun* (18 April 1939).

81 NIDS 1-391 *Sentō shōhō senji nisshi* 146: Dai-5 Suirai Sentai (December 1941).

82 Asō, *Rinkō*, 115; Rasa Kōgyō Kabushiki Gaisha Shashi Henshūshitsu, *Rasa Kōgyō 80-nen shi,* 114–15.

5 Disaster

The Abandonment of Japan's Remote Islands, 1902–1945

Even before the war there were signs that many remote island settlements were struggling. The most dramatic example of this occurred on Torishima, Japan's first bird island colony, which erupted in the summer of 1902. The 125 people on the island at the time all perished.[1] A second attempt to settle the island was abandoned in 1922, likely due to the dwindling revenue to be had from plumage. In 1926 the Hachijōjima local government sponsored a third and final attempt to settle the island, this time as a cattle ranch.[2] The colony limped along but, by 1939, was on the brink of insolvency by the time Torishima began to rumble again.[3][4] This time the island's twenty-three adults, five children and forty-five head of cattle were evacuated safely, but the subsequent eruption completely transformed the topography of the island, forming a new caldera on top of what had been the main harbour and settlement (Figure 5.1).[5] This more or less put paid to any plans to resettle Torishima, even had there been any people willing to commit years of their life to building a community on an island that had erupted twice within living memory, and might very well erupt again at any moment.

Torishima was an extreme example, in which the very island itself repulsed human efforts at settlement. But seven other Japanese islands were abandoned between 1920 and 1945. Mizutani Shinroku's colony on Marcus Island was abandoned in 1920, and Tsunetō Noritaka's mine on

1 'Minami-torishima jiken no kaiketsu', *Asahi Shimbun* (6 September 1902).
2 Okuyama Shūsaku, 'Zasshu zaisan keizoku kashisage mōshikomisho' (24 December 1937) [Original housed in TMHB, copy provided to author by Hasegawa Hiroshi].
3 Satō Kazuichi, 'Torishima keiei ni kan suru fukumeisho' (28 November 1937) [Original housed in the TMHB, copy provided to author by Hasegawa Hiroshi].
4 '"Kimi yo shiru – minami no umi no paradaisu 'Ahōdori no Torishima"', *Nankai Taimusu* (13 March 1939).
5 'Torishima kyū funkakō bakuhatsu! Hatsuka zen dekasegibito munen no hikiage', *Nankai Taimusu* (23 August 1939).

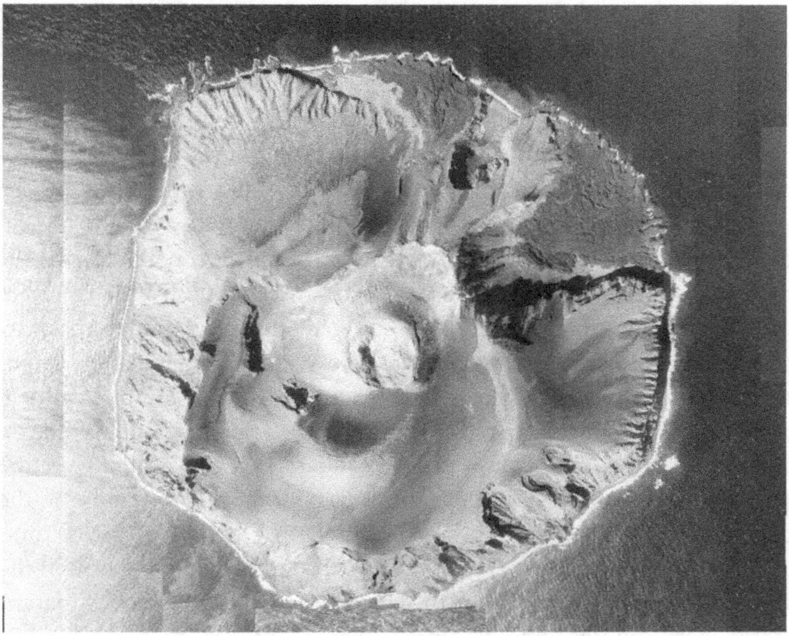

Figure 5.1 Satellite image of Torishima (1978)
© National Land Image Information, Ministry of Land, Infrastructure, Transport and Tourism

Rasa ceased operation for the second time in 1936.[6] Koga Tatsushirō abandoned his bonito processing plant on the Senkaku Islands in 1940. And although the two largest Bonin Islands of Chichijima (formerly Peel Island) and Hahajima remain inhabited to this day, colonies on five smaller islands were abandoned during the interwar period.[7] (See Appendix.)

The outbreak of the Pacific War heaped death and devastation upon those few remote island communities that remained. By the end of 1942 it had become clear that the conflict would be fought on an island-by-island basis across the Pacific, as American forces scrabbled to gain a purchase on any scrap of land from which they might launch aerial bombing raids on the Japanese mainland. As it was not at all clear which route the US

[6] Asō, *Rinkō*, 60.
[7] Enosawa Yukihiro, *Ritō to hō: Izu Shotō Ogasawara Shotō kara kenpō mondai o kangaeru* (Tokyo: Hōritsu Bunkasha, 2018), 117.

would pursue across the Pacific, the Japanese Navy billeted at least a token detachment upon even the most apparently insignificant of islands – either to keep watch for enemy forces or to establish meteorological facilities that would aid navigation.[8]

One such detachment was posted to Rasa Island, where Tsunetō's company had restarted operations in order to supply phosphate for the war effort. Though the soldiers got the Okinawan mine workers to help dig rudimentary air raid shelters, they were for the most part at the mercy of the enemy. The garrison subsisted on locally caught fish and vegetables grown in the company farm, and suffered periodic bombardment by planes and submarines. Boredom was also acute – although at one point their stay was enlivened by a visiting troupe of Korean 'comfort women' (forced sex workers). One soldier later recounted that he spent his time on Rasa listening to news of successive military defeats coming over the wireless, and waiting for inevitable martyrdom at the hands of the enemy.[9]

The war utterly transformed the lives of island civilians as well. If the Pacific War was a 'total' war, in which the populations of combatant states were mobilised to an unprecedented degree, then on Japan's remote islands it was particularly total. The Bonins, for instance, were transformed into a communications and logistical base for the Navy. The islands were placed under martial law, and as many as 25,000 troops were stationed on the main island of Chichijima, outnumbering the island's civilian population by more than three to one. Descendants of Pacific Islander and beachcomber settlers came under intense suspicion as potential fifth-columnists, and anyone discovered speaking English was called in for (sometimes brutal) interrogation by the secret police. This situation only worsened as Japan's war situation deteriorated. By 1944 Allied submarines had thrown Japan's shipping lanes into chaos, effectively cutting remote islands off from mainland supply lines. Hunger and malnutrition were the common experiences of almost all Japanese during the Pacific War, not to mention those non-Japanese who had the misfortune to find themselves under

8 After the war many of these meteorological stations would be inherited by Japan's civilian meteorological agency, and to this day they help monitor the path of incoming typhoons. See Fujisawa Tōru, *Ahōdori* (Tokyo: Tōkō Shoin, 1967), 29.

9 Morita Yoshio, *Rasa-tō shubitai ki: gyokusai o kakugo shita heishitacji no ningen dorama* (Tokyo: Kōjinsha NF Bunko, 1995), 118, 195.

wartime occupation. But it was felt particularly acutely on remote islands such as the Bonins. Eventually the bulk of the civilian population was evacuated to the mainland, while the few men who had not already been drafted into the military were ordered to stay behind and grow food for the garrison.[10]

Then there were those unlucky few islands that found themselves thrust into the front line of the conflict, their earth scorched by bitterly fought battles as US marines island-hopped across the Pacific theatre. Iwo Jima was in most respects a typical bird island: surveyed in the late nineteenth century with Japanese government support, settled during the first half of the twentieth century and more or less uninhabited since the war. What makes it distinctive are the sulphur reserves (its name literally means 'sulphur island') that issue from its volcanically active caldera. Well before the commercial value of offshore phosphate became apparent, sulphur was in high demand for matchmaking, and a mine had begun operating on the island in 1892. But as the sulphur reserves depleted the mining company diversified into cash crop agriculture, and by the Pacific War a thousand or so tenant farmers recruited from Hachijōjima and the Bonins were scratching a living on the island, coaxing sugar cane, coca and lemongrass from its ashen soil.[11]

Then, in 1944, the US military high command decided to launch a ground invasion of Iwo Jima, which made an ideal base from which to launch bombing raids on mainland Japan. In the run-up to the invasion the US subjected to the island to three days of close-range aerial bombardment before launching an amphibious assault by 70,000 marines in February 1945.[12] It produced a staggering casualty toll. An estimated 6,821 Americans died during the invasion, more than in any other battle in US history.[13] The death toll on the Japanese side was far higher. Of the 22,000 garrisoned to defend the island, just over 1,000 survived.[14] Most of the civilian population of Iwo Jima was

[10] Eldridge, *Iwo Jima and the Bonin Islands in US–Japan Relations*, 35–6, 104–5.
[11] Ishihara, *Iōtō*, 32, 41–2. Much of the coca harvested on Iwo Jima ended up being processed into cocaine and sold on the black market in India and Southeast Asia. See also Steven B. Karch, 'Japan and the Cocaine Industry of Southeast Asia, 1864-1944', in Paul Gootenberg, ed., *Cocaine: Global Histories* (London: Routledge, 1999) 146–61.
[12] Eldridge, *Iwo Jima and the Bonin Islands in US–Japan Relations*, 52, 64.
[13] Ishihara, *Iōtō*, 125.
[14] Eldridge, *Iwo Jima and the Bonin Islands in US–Japan Relations*, 68, 86.

evacuated to mainland Japan before the invasion began, but a hundred-odd islanders were ordered to stay on the island to provide logistical support: cooking and cleaning, preparing food and water, and acting as guides. Of these, only ten survived.[15]

Indigenous Perspectives on the Pacific War

Before we move on to the post-war period I want to spend a few paragraphs discussing the history of Micronesia under Japanese rule. Unlike bird islands, which had hardly any history of human habitation before they were annexed by Japan, the islands of Micronesia have been settled by Pacific Islanders for thousands of years. In this sense they sit somewhat outside the organising framework of the book. And yet their history is worth touching on for several reasons.

Firstly, the history of Japan's Pacific Empire is incomprehensible without understanding the central strategic, symbolic and economic role of the South Seas Mandate, set up after Japan invaded Germany's Pacific colonies during World War I. Micronesian islands formed part of the Empire of Japan's outer defensive perimeter (Figure 4.3), while also symbolising its status as a benevolent imperial power and responsible member of the international community. Remarkably, even after Japan stormed out of the League of Nations in 1933, it continued to submit reports on the progress of the South Seas Mandate to Geneva until as late as 1938.[16]

Secondly, the history of bird islands frequently interlinks with that of the Indigenous Pacific, even if only indirectly. We have already seen how bird island prospectors such as Mizutani Shinroku cut their teeth trading with Germany and Spain's South Pacific colonies, and how others such as Nishizawa Yoshiji served as advisers to the Imperial Japanese Navy. Micronesia loomed large in other ways, too. The vast mine on the Micronesian island of Angaur, whose management Japan inherited from the German Deutsche Sudseephosphat A.G, in many years produced more phosphate than all other Japanese phosphate mines combined.[17] The vast amount of Angaur phosphate that flooded

[15] Ishihara Shun, *Iōtō*, 131–2.

[16] Thomas W. Burkman, *Japan and the League of Nations: Empire and World Order, 1914–1938* (Honolulu: University of Hawai'i Press, 2008), 197–8.

[17] Peattie, *Nan'yō*, 67; Kumazawa, 'Tsunetō Noritaka to Nihon no rinsan shigen', 58–9.

the Japanese market in the 1920s was one of the factors that helped drive the Rasa Company out of business.

Finally, oral testimony from Indigenous Micronesians offers a vital alternative perspective on the rise and fall of Japan's empire in the Pacific, for they allow us to draw comparisons between islander-colonist relations in Micronesia and Japan's other Pacific territories such as the Bonin Islands and Iwo Jima. No two islands are alike, of course, so it is dangerous to extrapolate. But here I will focus in particular on the island of Saipan (Chamorro: Sa'ipan), one of the most populous Micronesian islands and also one that saw some of the deadliest fighting of the Pacific War.

Saipan's Indigenous Chamorro population has resided on Saipan since pre-Columbian times, but by the nineteenth century had been under Spanish rule for centuries. After the 1898 Spanish-American War Spain sold the islands to Germany, which in turn lost them to Japan in 1914. During the interwar period, lavish subsidies allocated by the Imperial Japanese Navy had helped transform the island into a sugar plantation worked primarily by migrants from Okinawa, Hachijōjima and Korea, so that by 1942 migrants from elsewhere in the Japanese Empire outnumbered Indigenous Chamorro by more than nine to one.[18]

Not every Chamorro account of Japanese colonial rule was negative. Some islanders earned a comfortable income leasing their land to Japanese-run commercial enterprises, and others found work in construction or performing menial roles within the colonial administration.[19] Still, Japanese racial chauvinism was pervasive. Chamorro children were schooled separately from their 'Japanese' counterparts, and there was very little effort to provide higher education or opportunities for career advancement. And local regulations enforced a petty kind of three-tier apartheid on the island: for instance, not only Chamorro but also Koreans were prohibited from drinking liquor.[20] Other Chamorro islanders recalled brutal regimes of corporal

[18] Ohara Tomoko, 'Sengo Okinawa shakai to Nanyō Guntō Hikiagesha – hikiagesha dantai katsudō ni chūmoku shite', *Imin Kenkyū* 6 (March 2010): 23; Tsushima Hideko, 'Kita Mariana Shotō kara no hikiagesha: Hachijōjima tōmin no imin jirei o chūshin toshite', *Hakusan Jinruigaku* 11 (March 2008): 152; Ishihara Shun, *Iōtō*, 67.

[19] Bruce M. Petty, *Saipan: Oral Histories of the Pacific War* (Jefferson: McFarland & Co., 2009), 30–1.

[20] Petty, *Saipan*, 35.

punishment. Felipe Iguel Ruak remembered how her friend's eyelashes were cut off after the friend was caught talking in class, as well as other, more extreme punishments that led to serious injury or even death.[21]

During the Pacific War, Japanese rule became considerably harsher. As the prospect of a US invasion loomed, men, women and even children were forced to perform unpaid labour expanding Saipan's airfield.[22] Some Chamorro men who spoke Japanese well enough were sent to help interrogate prisoners of war on the neighbouring American colony of Guam, creating a legacy of intra-Chamorro bitterness that lingered long into the post-war period.[23] At the same time, the Chamorros' racial Otherness made them intrinsically suspicious, and the secret police began conducting night-time raids, ransacking homes in search of wireless receivers that, they suspected, were being used to communicate with the enemy. Acculturation programmes were ramped up: churches were closed, and the Chamorro language was banned.[24]

This racialised thinking would also have profound consequences during the battle itself. On Iwo Jima the majority of the civilian population (except for military-age males in good health) was evacuated to safety in advance of the battle, albeit taking only what they could carry. Bonin Island residents, even descendants of the original settlers, were evacuated in a similar fashion.[25] On Saipan, however, the authorities decided to evacuate only those civilians who had migrated to the island from elsewhere in the Japanese Empire; the Chamorro by contrast were essentially left to fend for themselves.[26] The decision may have made sense according to the racially hierarchical logic of the Japanese Empire, which posited Hachijōjimans, Okinawans and Koreans as somehow more 'Japanese' than Indigenous Micronesians. But in reality few of the evacuees would ever have lived in mainland Japan before.

21 Stephanie Soder and Jennifer F. McKinnon, *It Rained Fire: Oral Histories from the Battle of Saipan* (Lulu.com, 2019), 3–5.

22 Petty, *Saipan*, 25.

23 Keith L. Camacho, *Sacred Men: Law, Torture, and Retribution in Guam* (Durham, NC: Duke University Press, 2019); Petty, *Saipan*, 26.

24 Petty, *Saipan*, 34.

25 Chapman, *The Bonin Islanders*, 139–40; Eldridge, *Iwo Jima and the Bonin Islands in US–Japan Relations*, 34.

26 Tsushima, 'Kita Mariana Shotō kara no hikiagesha', 151.

The departed 'Japanese' colonists were replaced by an influx of 30,000 soldiers to garrison Saipan against the incoming American invasion.[27] Many Chamorro were turfed out of their homes in order to billet the incoming troops. Some went to stay with relatives, but others were not so lucky: Thomasa Naraja, for instance, recalled her parents taking her and her twelve siblings to live in a cave.[28] Things became even worse when the invasion began. Escolita Tudela Cabrera remembered that 'when the fighting started the Japanese became mean. We became like slaves – forced labor. Then the Japanese really treated us like third-class citizens. They would come to our farm and take whatever they wanted. If we protested they would threaten us.'[29]

Nearly a quarter of Saipan's Chamorro population died during the battle itself, and survivors told horrific tales of cramming into caves for weeks on end, surrendering only after their water ran out.[30] Vicky Vaughan, eleven at the time, recalled being driven from her family's farm by artillery shelling only to get caught up in fighting between Japanese and American troops. She saw her sister disintegrate before her eyes, her baby brother's head smashed in, her cousin 'trying to push her intestines back into her stomach with her dirty hands'. Her step-mother and seven of her nine siblings died during the attack, and Vicky herself had her back scorched by a flamethrower while she huddled in a foxhole. Afterwards US soldiers sent her to live in a squalid refugee camp (Figure 5.2), where she and the other survivors were introduced to the delights of spam.[31]

The Aftermath

World War II transformed Japan, East Asia and the Pacific in profound ways. Alongside lives, landscapes and property, the war also destroyed empires – not least Japan's own. Between 1945 and 1952, Japan itself

[27] Petty, *Saipan*, 7. [28] Soder, *It Rained Fire*, 6. [29] Petty, *Saipan*, 29.

[30] Matthew Hughes, 'When Soldiers Kill Civilians', *History Today* 60 no. 2 (27 January 2010); John F.; Embree and Arthur C. Huston, 'Military Government in Saipan and Tinian: A Report on the Organization of Susupe and Churo, Together with Notes on the Attitudes of the People Involved', 5 no. 1 (1946): 5; Petty, *Saipan*, 30.

[31] Petty, *Saipan*, 20–1. For the relationship between militarism, empire and diet in the American Pacific see Mire Koikari, 'Food, Military and Empire in Post-World War II Okinawa', in *Devouring Japan: Global Perspectives on Japanese Culinary History*, ed. Nancy K. Stalker (Oxford: Oxford University Press, 2018): 171–83.

(a)

(b)

Figure 5.2a A Chamorro family interned at Charon Kanoa, Saipan (July 1944)
Source: National Archives and Records Administration digital collection
Figure 5.2b 'Terechan Kiyatake, a Japanese nurse working in a US military hospital on Saipan, chats with Terumena Pinkerinon, a Chamorro who is a graduate of Tokyo' (July 1944)
Source: National Archives and Records Administration digital collection

was occupied by a US-dominated coalition of Allied powers who used the opportunity to redraw Japan's borders at will. China reclaimed Manchuria and Taiwan, and the US and the Soviet Union each set up their own client state on the Korean peninsula. Japan's oceanic boundaries were also revised. Sakhalin and the Kurile Islands were occupied by the Soviet Union, and at the time of writing remain firmly under Russian control.[32] The United States detached not only the South Seas Mandate but also the Bonin, Amami and Ryūkyū Islands from Japanese rule. Even the Izu Islands briefly came under foreign occupation, for in January 1946 the Supreme Commander Allied Powers (SCAP) declared the islands separate from the Japanese mainland. The decision was reversed after two months, but not before the ensuing confusion inspired one village headman to suggest that his island of Ōshima, having 'lost its ancestral land', draft its own constitution as an independent nation.[33]

The US government was not sure what to do with these islands at first. Many had been captured from Japan at the cost of immense loss of life. Moreover, their strategic value as 'unsinkable aircraft carriers' had not lessened in the atomic age: they made for discreet places to store and even test nuclear weapons at a remove from electoral scrutiny.[34] During the war, there had been talk of 'returning' the Ryukyu Islands to China, and in August 1945 the House Committee of Naval Affairs published a report that suggested rolling all Japan's outlying and mandated islands together to create the fifty-first American state, to be known as 'The State of the American Pacific'.[35] This wild-eyed scheme came to nothing, and instead the former South Seas Mandate was relabelled the United Nations Trust Territory of the Pacific Islands under the trusteeship of the United States – the fourth colonial power to rule the island group in half a century. Most of the islands remain de facto US colonies to this day.

32 Bukh, *These Islands Are Ours,* Ch.1.

33 Izu Shotō Tōkyō Ikan 100-nen Shi Hensan Iinkai, *Izu Shotō Tōkyō ikan 100-nen shi*, 759.

34 For the phrase 'unsinkable aircraft carrier' see John G. Morris, 'The Air Assault on Japan', *Flying Magazine* (November 1943). The phrase would later be repeated by Prime Minister Nakasone Yasuhiro to refer to Japan's geopolitical situation during the Cold War.

35 Kimie Hara, 'Okinawa, Taiwan, and the Senkaku/Diaoyu Islands in United States–Japan–China Relations', *Asia-Pacific Journal: Japan Focus* 113 no. 28 (July 2015): 4; Immerwahr, *How to Hide an Empire*, 228.

As for the other islands, Washington did eventually recognise what it termed Japan's 'residual sovereignty' over the Ryukyus and Bonins, but exact negotiations as to the timing of their return to Japanese administration continued until the late 1960s, and until then both island groups remained under indefinite US occupation.[36] Echoing the racialised logic of their vanquished enemies, the American administrators refused to allow 'Japanese' evacuees from Saipan, Iwo Jima or the Bonins to return to their homes on the islands. Adroit petitioning by the great-grandson of Nathaniel Savory did, however, convince SCAP to let only 'islanders of Western descent' return to the Bonins. Distinguishing 'Western' from 'Japanese' Bonin Islanders was not straightforward: the US consul to Japan noted that 'Most were third-generation Boninites who had intermarried with Japanese, spoke English poorly if at all, and had tenuous ties with the United States.' The task was deputised to the International Red Cross, which seems to have substantially relied on a list supplied by Savory. Eventually 125 islanders, including quite a few 'Japanese' wives, were allowed to return to the Bonins.[37] The rest had to wait until the islands reverted to Japanese rule in 1968. The evacuated colonists of Iwo Jima were even less lucky, for they were unable to return to the island even after it reverted to Japanese sovereignty. Today the island is a Japanese airforce base.[38]

Japan's Remote Island Problem

Meanwhile, those remote islands that did remain under Japanese control were plunged into a state of acute economic crisis after the war. Hachijōjima illustrates their plight particularly well. We first met the islanders of Hachijōjima at the very start of our story, brawling aboard a boat off the Bonin Islands. In the years that followed, Hachijōjimans had been at the forefront of efforts to colonise Japan's oceanic empire. It was Hachijōjimans, for instance, who had worked as overseers on Rasa Island, taken the lead in colonising Micronesian islands such as Saipan and Tinian, and followed the Japanese Navy as it pressed further on into the South Pacific during World War II.[39] Some died in the attempt: an official history of the island reported that many

[36] Eldridge, *Iwo Jima and the Bonin Islands in US–Japan Relations*, 161.
[37] *Ibid*, 194–5. [38] *Ibid*, 68.
[39] Tsunetō, *Discovery and Exploitation of Rasa Island*, 15, 20; Tsushima, 'Kita Mariana Shotō kara no hikiagesha', 149.

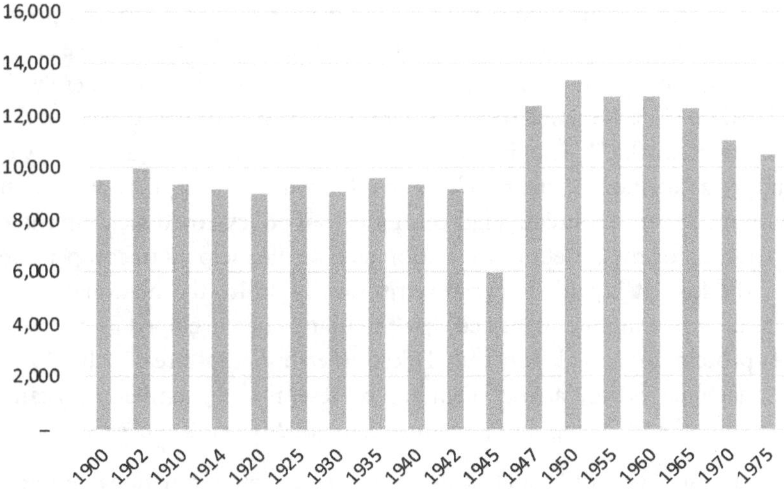

Figure 5.3 Hachijōjima population, 1900–1975
Source: *Izu Shotō Tōkyō ikan 100-nen shi* (1981)

Hachijōjimans 'chose to commit suicide in the face of the enemy's advance'.[40]

Thanks to these efforts, those islanders who stayed behind on Hachijōjima had been sustained in large part by remittances.[41] But after Japan's defeat, the Hachijōjiman colonists were abruptly repatriated, and between 1945 and 1947 the island's population increased to more than a third above its prewar level (Figure 5.3).[42] Moreover, the repatriates returned to an island economy that had been devastated by the war. Soldiers billeted on the island had slaughtered its cattle herds for food, denuded its hillsides for firewood and eventually burned its silk looms as well. Meanwhile, deserted fields had run wild, while irrigation facilities fell into disrepair. The result of all this was to put acute pressure on resources, with later returnees unable to rent agricultural land or even to salvage materials to build shelter for themselves.[43]

Even after initial shortages had been dealt with, the government was still faced with the problem of how to sustain these outlying islands as

[40] Izu Shotō Tōkyō Ikan 100-nen Shi Hensan Iinkai, *Izu Shotō Tōkyō ikan 100-nen shi*, 1073–4.
[41] *Ibid*, 348–9, 1072. [42] *Ibid*, 1067–8. [43] *Ibid*, 347.

viable economic units, now that their residents had been shorn of opportunities for out-migration to the Japanese Empire. The islanders' elected representatives were canny enough to couch their plight not merely as a local one but as related to a broader problem of national survival in the post-war world.[44] The islands' desperate poverty, they argued, was simply a function of their lack of development; in fact, they were rich with untapped resources. Vigorous industrial policy could yield up these resources for the benefit of the nation as a whole, helping to compensate for the loss of Japan's imperial territories. In 1948, the National Land Planning Committee suggested that the islands be developed as a source of protein and fuel for the Tokyo metropolitan area.[45] In 1953 Tsunashima Seikō, a Diet member representing a number of island communities in Nagasaki Prefecture, reminded his colleagues that:

> Our ancestral land of Japan has lost forty-five percent of its national territory as a result of losing the war . . . In order to achieve political independence and economic self-sufficiency for our nation on this limited amount of territory with its enormous population of over 80 million, then along with promoting trade it is also urgent to plan for the development of every last inch of national territory. The remote islands separated by sea from the Japanese mainland harbour many marine and terrestrial resources, but due to restrictions imposed by various natural and social conditions, they are mostly as yet undeveloped.[46]

Remote islanders also found unexpected allies among Japan's community of anthropologists and ethnologists. Before 1945 many of these scholars had been accustomed to conducting fieldwork in the colonies: Yanagita Kunio, the doyen of folklore studies in Japan, had even developed an elaborate 'Southern Islands Theory' that traced the origins of the Japanese people to Okinawa, Taiwan and Japan's South Seas Mandate islands in Micronesia.[47] The loss of empire, and the Occupation-era restrictions on overseas travel, forced Japanese ethnographers to recalibrate their search for the soul of traditional Japan.[48]

44 The Hokkaidō Prefectural Government made a similar argument when lobbying for the return of the Kurile Islands. See Bukh, *These Islands Are Ours,* 41.
45 Speech by Nagai Katsujirō to 2nd National Diet, National Land Planning Committee (Kokudo Keikaku Iinkai) No.17 (3 July 1948).
46 Speech by Tsunashima Seikō to 16th National Diet, Economic Stabilization Committee (Keizai Antei Iinkai) No.4 (7 July 1953).
47 Murai Osamu, *Nantō ideorogii no hassei: Yanagita Kunio to shokuminshugi* (Fukutake Shoten, 1992); Tierney, *Tropics of Savagery*, 81.
48 Sakano Tōru, 'Kyū Shakai Rengō no kyōdō chōsa to "kokudo"', *Nihon Daigaku Keizai Gakubu Sangyō Keiei Kenkyūjō Shohō* 263 (December 2011): 1;

A 1950 expedition to the island of Tsushima, lying in the strait between mainland Japan and Korea, helped set the tone of much post-war Japanese anthropological research, and one of the expedition's more eminent alumni, Miyamoto, Tsuneichi, went on to play a key role lobbying the Japanese government on behalf of island communities.[49]

Partly thanks to efforts such as these, in July 1953 the National Diet passed the Remote Islands Development Law, a ten-year package of ¥2.7 million in central government funding earmarked for investment in remote islands infrastructure such as fisheries facilities, shipping lane beacons, harbour improvement, road construction, electrification, afforestation and shoreline protection, erosion control, education and sewerage.[50] The law remains on the books today and is the basis for a series of rolling ten-year aid packages designed to subsidise island economies.[51] It has without question provided a crucial lifeline to communities on islands such as Hachijōjima, helping to at least slow the pace of depopulation. Still, in the years since World War II another fifty-eight Japanese islands have been abandoned, some of which (like Hachijōjima's sister island of Hachijō-kojima) had been settled for hundreds of years.[52] (See Appendix.)

In the immediate post-war period, then, Japan's remote islands had transformed from stepping-stones of colonial expansion to sites of trauma and anxiety – about territorial loss, demographic decline and the vanishing of tradition. It was within this context that new ways of

Eiji Oguma, *A Genealogy of 'Japanese' Self-Images* (Tokyo: Trans Pacific Press, 2002): 299.

49 Tsukishima Kenzō, 'Tsushima, Amami to tōji no seiji jōkyō to no kakuwari ni tsuite: chōsa no kaki', *Jinrui Kagaku* 16 (Kaimeidō: 1964): 131; Sakano, 'Kyū Shakai Rengō no kyōdō chōsa to "kokudo"', 4, 13; Tsuneichi Miyamoto, *The Forgotten Japanese: Encounters with Rural Life and Folklore* (Berkeley: Stone Bridge, 2010), 9; Miyamoto Tsuneichi, 'Shima no bunka' (December 1953) in *Miyamoto Tsuneichi ritō ronshū vol. 1* (2009); Yuji Ankei, 'The Fieldwork Ethics and Positionality of Tsuneichi Miyamoto: A Pioneer of Border Island Studies in Japan', *Eurasia Border Review* 4 no. 1 (2013): 25, 31–2; Tsuneichi Miyamoto, *Ritō ronshū* (Tokyo: Mizunowa Shuppan, 2009). For more on Miyamoto, see Alan S. Christy, *A Discipline on Foot: Inventing Japanese Native Ethnography, 1910–1945* (Lanham: Rowman & Littlefield, 2012), 30, 187.

50 TMA 18.84 *to* 30, 'Shōwa 30 nendo ritō shinkō keikaku jigyō'.

51 Kokudō Kōtsūshō, 'Ritō Shinkō' [n.d.]: www.mlit.go.jp/kokudoseisaku/chirit/kokudoseisaku_chirit_fr_000003.html. Last accessed 22 June 2020.

52 Enosawa Yukihiro, *Ritō to hō: Izu Shotō Ogasawara Shotō kara kenpō mondai o kangaeru*, 116–17.

thinking about deserted islands began to emerge – not only as sites for economic development, but also for the conservation of valuable but threatened nature. Witness, for example, post-war efforts to promote tourism to the Izu Islands by preserving them as 'a beautiful volcanic island chain that can be made to represent Japan.'[53] This same context also informed a newfound interest in the birds that nested on remote islands – now viewed not as resources to be extracted but as treasures to be guarded. It is this story that we shall explore in Chapter 6.

[53] Kikuchi Jinzō, 'Izu Shotō ni okeru hito to sono kankyō' in Tōkyō-to', *Izu Shichitō gakujutsu chōsa hōkoku* (March 1957): 112.

6 *Resurrecting the Torishima Albatross*

Wild Birds and Sovereignty in Post-War Japan

In the Anglo imagination, albatrosses have been 'charismatic megafauna' since before the term was invented. When encountered during voyages of colonial exploration into the southern hemisphere, they were construed by anxious British sailors as omens of good luck.[1] They were eventually immortalised through the stanzas of Samuel Taylor Coleridge's *Rime of the Ancient Mariner.* Coleridge's poem, reprinted in at least twenty editions over the course of the nineteenth century, freighted the albatross with a range of moral and spiritual associations, transforming it into the original avatar of the wilderness: a creature whose ill-treatment provoked the wrath of that 'tutelary spirit ... who bideth by himself / In the land of mist and snow', and so brought judgement raining down on man.[2]

Though Anglo writers may have romanticised albatrosses, Japanese attitudes to the birds were brusquely unsentimental. Indeed, the Japanese name for Steller's albatross translates simply as 'stupid bird' – a reference to its ungainly waddle while land-bound and initial lack of flight reflex when confronted with humans.[3] The etymology has more than a passing similarity to that of the Mauritian dodo (from Portuguese, *doudo* meaning simpleton) – another large, island-dwelling bird. And as we have seen, Japanese hunters took advantage of such evolutionary traits to slaughter the species on its nesting grounds on

[1] Bernard Smith, *Imagining the Pacific: In the Wake of the Cook Voyages* (New Haven: Yale University Press, 1992), 154–5; Brooke, *Albatrosses and Petrels*, 4; Arthur Frederick Gotch, *Birds – Their Latin Names Explained* (London: Blandford Press, 1981), 190. Even if these beliefs were held with any seriousness, however, they did not prevent sailors from regularly killing albatrosses for food or sport.

[2] William Wordsworth and Samuel Taylor Coleridge, *Lyrical Ballads: 1798 and 1802* (Oxford: Oxford University Press, 2013): 5–24.

[3] Today albatrosses are most commonly known as *ahōdori* but other, older names for the bird (*bakadori, tōkurō*) have similar meanings. See Sugawara Hiroshi and Kakizawa Ryōzō, *Zusetsu chōmei no yurai jiten* (Tokyo: Kashiwa Shobō, 2005).

islands throughout the North Pacific. By the eve of World War II it seemed likely that Steller's albatross would also go the way of the dodo.

But in the post-war period Japanese attitudes to albatrosses changed utterly. The birds became the object of a sustained conservation campaign that began immediately after the end of the US occupation of Japan, so that by 1958 Japan's Agency for Cultural Affairs would declare the albatross nesting grounds on Torishima to be a 'natural monument' of the nation. The same year the International Union for the Conservation of Nature added Steller's albatross (*Diomedea albatrus*) to its Lake Success List, a list of endangered species that would eventually evolve into the Red List, the world's most authoritative biodiversity database. These linked declarations were accompanied by a nationwide campaign to raise public awareness of the plight of wild birds. Front-page editorials in national newspapers cherished Torishima as a bird 'paradise', and sermonised about the importance of fostering love of nature in the hearts of the populace.[4] And the post office chose to commemorate the species by emblazoning it upon a 20-sen stamp (Figure 6.1).

How should we fathom this sea-change in Japanese attitudes to the albatross? Conservationists have generally framed it as part of a trajectory of national moral renewal, in which the dark era of Tamaoki's state-sanctioned albatross culls gave way to a post-war commitment to save the species from the brink of extinction. There is no reason to doubt the sincerity of Japanese ornithologists' commitment to this cause, which for many amounted to a lifelong passion. Nevertheless, framing the story simply as a morality tale omits the class politics that often characterise wildlife conservation in practice. It also overlooks a post-colonial context that was distinctive to post-war Japan. For one thing, by 1945 Japan's bird island colonies had all been abandoned, rendering the opportunity cost of conservation strikingly low. But bird conservationists also consciously fashioned the fall and rise of Torishima's albatrosses as an allegory for mid-twentieth century international relations, hoping to exploit ambient anxiety about Japan's compromised sovereignty following defeat in World War II.

4 'Aka-chan mo maruba genkina tori suna bi ahōdori', *Asahi Shimbun* (19 January 1961); 'Ahōdori no nakama (aichō shūkan ni yosete)', *Asahi Shimbun* (7 May 1961).

Figure 6.1 Stamp issued by Japan Post as part of a 'Nature Conservation Series' (6 January 1975)

Even more so than their British and American counterparts, Japanese ornithologists tended to hail from elite backgrounds. This was partly because ornithology was for much of its history the preserve of amateurs, and the discovery of new taxa required a great deal of spare time, as well as the financial wherewithal to mount costly expeditions to remote parts of the world. But in addition, ornithology in Japan is unusual for being historically dominated not just by the wealthy but by men of aristocratic lineage. Studying and advocating for the natural world provided such men with a means to carve out a new role in public life at a time when their hereditary privileges were being abolished. It enabled them to accumulate new reserves of symbolic capital even as old ones were being eroded.

Secondly, the campaign to save the Torishima albatross echoed shifting trans-national attitudes to wildlife management on the part of states and the scientists who advised them. In fact, the very notion of using commitment to nature conservation as an index of a nation's civilisational level was imported during this period from the United States. The circumstances under which this notion was introduced to Japan were particular to the nation's linked experiences of defeat, occupation and territorial amputation after World War II. The Allied Occupation provided American ornithologists with an unparalleled opportunity to impose on their vanquished enemy a version of their own thinking about how people should relate to the natural world. Their Japanese counterparts seized on this opportunity eagerly, espousing wildlife conservation to the public as a way for their country to demonstrate its reintegration into the post-war international community.

Wild Birds and National Treasures

The killing and conserving of birds has a long history in Japan. Commercial hunters had operated widely during the Tokugawa period, catering to a brisk demand for bird flesh. (One Edo cookbook ranked bird meats from tastiest to least palatable, beginning with crane, followed by swan, wild goose, wild duck, and finally chicken.)[5] But they had generally done so with the permission of the village-level authorities that governed commons usage, offering their services as eradicators of pests and dangerous animals in exchange for permission to hunt game in the vicinity.[6] But political reforms during the Meiji Period (1868–1912) transformed Japanese hunting practices, stimulating a dramatic increase in the number of birds killed for commercial purposes. Partly this was due to the de facto nationalisation of forestland by the Meiji state. The central government's decision to designate formerly locally-administered land as royal forest demolished the relationships between hunters and local authorities, while failing to effectively deter hunters from poaching on the newly nationalised land.[7] The

[5] Haruo Shirane, *Japan and the Culture of the Four Seasons: Nature, Literature, and the Arts* (New York: Columbia University Press, 2011), 118.

[6] For a discussion of hunters in Tokugawa Japan, see Taguchi Hiromi, 'Rettō kaitaku to shuryō no ayumi', *Tōhoku-gaku* 3 (2000); Taguchi Hiromi, 'Matagi – Nihon rettō ni okeru nōgyō no kakudai to shuryō no ayumi', *Chigaku Zasshi* 113 no. 2 (2004).

[7] Conrad Totman, *Japan's Imperial Forest: Goryōrin, 1889–1946: With a Supporting Study of the Kan/min Division of Woodland in Early Meiji Japan, 1871–76* (Honolulu: University of Hawai'i Press, 2007), 101.

result was to render bird-hunting a free-for-all activity at the very moment when the global market for decorative plumage (see Chapter 2) was beginning to surge in value.

The Meiji government was slow to respond to this decimation of Japan's bird population. The new government introduced the first game regulations for Japan as a whole in 1873, introducing a licensing system for hunting with guns. But this early legislation was intended less to conserve bird populations than to regulate gun ownership in the face of ongoing civil unrest – throughout the 1870s the young regime was preoccupied with fending off a series of armed regional rebellions. Tellingly, the 1873 regulations issued no restrictions on hunting with nets, despite this being the technique favoured by the overwhelming majority of hunters in Japan at the time.[8] It was not until 1892 that the Ministry of Agriculture and Commerce issued a more comprehensive set of regulations that required net-hunters as well as gun-hunters to purchase licenses.

These early attempts at conservation remained overwhelmingly instrumentalist, in the sense that they aimed at the sustainable management of populations for the benefit of hunting. At the turn of the twentieth century this began to change, however. As with Japan's energetic 1880s embrace of colonialism, the roots of this can be traced to a trans-national phenomenon, whereby nationalist thinkers embraced Late Romantic tropes in order to ground their imagined communities in environmentally determined notions of race and place: not in society, in other words, but in nature. In 1894, Shiga Shigetaka, whose massively successful *Conditions in the South Seas* had done so much to popularise *Nan'yō* expansion during the previous decade, published *On Japanese Landscape*, which celebrated 'the splendours and superiority of the Japanese landscape', while also urging the exploitation of its natural resources.[9] Shiga urged his readers to: 'climb a mountain, rinse your mouth in the pure crystalline waters of a crater lake, bite into snow that has fallen since time immemorial, shout into the wind, and raise your voice in song toward the expansive sky'.[10] The newly monikered 'Japan

8 Rinyachō, *Chōjū gyōsei no ayumi* (Tokyo: Rinya Kōsai-kai, 1979), 5–6.

9 Kären Wigen, '*Discovering the Japanese Alps: Meiji Mountaineering and the Quest for Geographical Enlightenment*', 31 no. 1 (2005): 12; Thomas R. H. Havens, *Parkscapes: Green Spaces in Modern Japan* (Honolulu: University of Hawai'i Press, 2011), 55–6.

10 Richard Okada, '"Landscape" and the Nation-State: A Reading of Nihon Fūkei Ron', in *New Directions in the Study of Meiji Japan*, ed. Helen Hardacre, Adam Lewis Kern (Leiden: Brill, 1997), 100.

Alps', transparently modelled on the Swiss Alps so beloved of Anglo Romantic poets, thereby furnished a Japanese equivalent to the English Lake District, the German Hercynian Forest or the American Rockies.[11]

Shiga never actually climbed any mountains himself, but he did succeed in inspiring a generation of hikers and amateur mountaineers, setting in motion a series of debates between geographers, naturalists, landscape architects, local tourism boosters and government agencies that culminated in the passage of the much more ambitious National Parks Law in 1931.[12] Another key driver of legislation in this period was Miyoshi Manabu, a botanist who had studied in Germany and been impressed by the measures undertaken by various state governments to designate and preserve cultural and natural heritage. On his return to Japan Miyoshi took up a professorship at Tokyo Imperial University, and in 1919 he successfully lobbied the National Diet to pass a Historical Remains, Scenic Place and Natural Monument Preservation Law based largely on German precedents.[13]

Following the passage of the law the Home Ministry set out to canonise the heritage of the nation. Many species designated as natural monuments had specific pre-existing cultural value, either at the local or the national level. Several breeding and migration grounds were set aside for cranes. Two different species of fighting cocks were singled out, as were birds such as the snow grouse that had historically been revered as messengers from the gods. The seagulls of Kabushima Shrine in Aomori were protected as emissaries of the deity Benzaiten, and to this day visitors who are spattered by birdshit (*fun*) are encouraged to buy a votive plaque to celebrate their good luck (*un*).[14] But some were

[11] Okada, '"Landscape" and the Nation-State', 7–8. Compare Shelley's 'The everlasting universe of things /Flows through the mind, and rolls its rapid waves . . .' with Shiga's 'if you contemplate deeply the silence that surrounds you, your mental powers will become that of a deity, you yourself will become sacred and gain awareness of the mysterious wisdom spontaneously arising out of you'. See Shelley, *Mont Blanc: Lines Written in the Vale of Chamouni* (1817); Okada, '"Landscape" and the Nation-State', 100.

[12] Havens, *Parkscapes,* 71–2.

[13] Uchida Seinosuke, *Tennen kinenbutsu: chōrui-hen* (Tokyo: Sōgensha, 1960), 9; Havens, *Parkscapes*, 62; Imort, 'A Sylvan People'; Natsuko Akagawa, *Heritage Conservation and Japan's Cultural Diplomacy: Heritage, National Identity and National Interest* (London: Routledge, 2014).

[14] Yamashina Yoshimaro. *Nihon no chorui to sono seitai* (Tokyo: Azusa Shobo, 1934): 1046–7: https://visithachinohe.com/spot/kabushima/. Last accessed 30 March 2022.

selected primarily for their rarity in purely Linnaean terms – what we would today call their biodiversity value. The first bird species awarded national monument status, for instance, was Lidth's Jay, found only on the Amami Islands (and a frequent target of commercial plumage hunters).[15] Unfortunately, natural monument status did not always help to protect the birds in practice. Indeed, in some cases it may even have been counter-productive: rumours abounded that some commercial poachers were even citing the 'monument' status of birds they captured as proof of their rarity (and therefore high market value).[16]

These mingled currents of Romantic aesthetics and nature conservation helped foster an appreciation of wild birds, in particular, in certain elite circles. In 1934, a number of ornithologists and prominent intellectuals founded the Wild Bird Society of Japan (*Nihon Yachō no Kai*) , an organisation dedicated to promulgating the idea that birds were best appreciated in their 'natural' environment, i.e. outside of captivity. The association regularly organised tours on which it taught participants to observe birds scientifically – which is to say using binoculars and field manuals. These tours also doubled as hunting expeditions for, like bird watchers elsewhere in the world at this time, the WBSJ membership generally saw no contradiction between appreciating birds and killing them. Despite this, though, it did develop a critique of *a certain kind of* bird slaughter that echoed Progressive-era Anglo bird conservationist movements: commercial hunting was to be deplored, but amateur hunting was legitimate insofar it had an ostensible scientific and/or educational purpose (see Chapter 3).[17]

Within a few years, the WBSJ's membership had grown to nearly 300, drawn from a particularly rarified stratum of Japanese society. Partly this was because of the dominant role played by aristocrats in the early development of ornithology in Japan: of the sixteen articles

15 BirdLife International, *Threatened Birds of Asia: the BirdLife International Red Data Book* (Cambridge: BirdLife International, 2001): http://datazone.birdlife.org/species/factsheet/amami-jay-garrulus-lidthi. Last accessed 30 March 2022.

16 International Committee for Bird Preservation, 'Report of the Japanese Delegation', *Bulletin of the International Committee for Bird Preservation* 2 (London: London International Committee for Bird Preservation, 1929): 38.

17 Setoguchi Akihisa, 'Yachō o meguru dōbutsu-kan', in *Nihon no dōbutsu-kan: hito to dōbutsu no kankeishi* (Tokyo: Tōkyō Daigaku Shuppankai, 2013), 166, 169.

published in the first ever edition of the Japanese ornithological journal *Tori* in 1915, seven were penned by men of aristocratic heritage.[18] But one man in particular, Yamashina Yoshimaro, deserves special attention. As the cousin (and childhood playmate) of the Shōwa emperor, Yamashina had been born a prince, before the 1907 reforms to the aristocracy demoted him to a mere marquis.[19] But whereas the young emperor cultivated an interest in marine biology, Yamashina instead developed a passion for birds.[20] As a boy he read (and annotated assiduously) Blakiston's *Birds of Japan*, and would later recall the joy he took in observing the geese gathered on the lawns of the Imperial Palace in the centre of Tokyo.[21] As a young man he enlisted as an officer in the Imperial Japanese Army, the expected career path for Japanese aristocrats. But his actual participation in military life was sporadic and frequently interrupted by leisure travel, including a great number of hunting trips and bird-collecting expeditions. In 1929 he dropped out of the army altogether, to enrol as an irregular student in the zoology department of Tokyo Imperial University. After two years of study he established his own private laboratory on family-owned land in Shibuya, Tokyo, in 1932. Later he would expand this into the Yamashina Institute for Ornithology, which would come to house a collection of more than 60,000 bird specimens.[22]

Though he was active in an age when the natural sciences were becoming increasingly professionalized, Yamashina's career as described so far bears a striking resemblance to that of many other noted ornithologists who had preceded him. There was a long tradition of gentleman amateurs parlaying their wealth and connections to achieve notability as taxonomists. The most famous of these was probably Joseph Banks, who had paid his way aboard Cook's expedition to the South Pacific and returned to instant celebrity. During the voyage Banks had exploited his

18 *Tori* 1:1 (1915). See also Isono Naohide, *Tonosama seibutsugaku keifu* (Tokyo: Kagaku Asahi Henshūbu, 1991).

19 Takie Sugiyama-Lebra, *Above the Clouds: Status Culture of the Modern Japanese Nobility* (Oakland: University of California Press, 1993), 59.

20 Aoki Eiji, *Yamashina Yoshimaro no shōgai* (Tokyo: Yamashina chōrui kenkyūjo, 1982), 20; Herbert P. Bix, *Hirohito and the Making of Modern Japan* (New York: Harper Collins, 2000), 60–1.

21 Yamashina Yoshimaro, *Tori no heru kuni fueru kuni: Ōbei tori angya* (Tokyo: Nihon Chōrui Hogo Renmei, 1967), 207–8.Yamashina's annotated copy of Blakiston is housed in the Yamashina Institute for Ornithology.

22 Aoki, *Yamashina Yoshimaro no shōgai*, 87.

position to take potshots at albatrosses from the deck of the *Endeavour*, pronouncing their flesh 'so good that every body commended and Eat heartily of them tho there was fresh pork upon the table'.[23] He returned to London bearing a brace of carcasses plus a series of life drawings that, amongst his many other scientific discoveries, established albatrosses as a distinct species.[24]

Yamashina wasn't merely wealthy like Banks. In his search for specimens that had hitherto escaped the taxonomical gaze, he also freely availed himself of colonial opportunities. In 1933 he embarked on a planned (though never completed) five-volume *A Natural History of Japanese Birds,* with the intention of producing a comprehensive reference work for all bird species found within the fast-expanding boundaries of the Japanese Empire. For the purposes of his project Yamashina divided Japan into three zones. The northern zone comprised the four 'inner' islands of Honshū, Shikoku, Kyūshū and Hokkaidō, but also the Kurile Islands (annexed 1875), the Bonin Islands (1876), the Volcano Islands (1891), Sakhalin (1905) and Korea (1910). There was also an Asian Zone that comprised the Amami Islands and the Ryūkyū Islands (annexed 1879) and Taiwan (1895), and an Australasian Zone, including the Marianas, the Carolines, the Marshall Islands and Palau, all of which had been seized by Japan during World War I and governed as a League of Nations Mandate (see Chapter 4).[25] The project self-consciously echoed John James Audubon's *Birds of America,* published a century earlier. Just as Audubon's magnum opus had sought to provide a taxonomical underpinning for the fast-expanding United States, so Yamashina catalogued the new biodiversity riches of Greater Japan, and in doing so constructed for the patchwork imperial polity a coherent basis in nature.[26]

23 Joseph Banks and J. C. Beaglehole, *The Endeavour Journal of Joseph Banks, 1768–1771* (Sydney: Trustees of the Public Library of New South Wales, 1962), 2: entry for 5 February 1769. Banks also added a brief recipe: 'The way of dressing them is thus: Skin them overnight and soak their carcasses in Salt water till morn, then parboil them and throw away the water, then stew them well with very little water and when sufficiently tender serve them up with Savoury sauce.'

24 Smith, *Imagining the Pacific*, 36.

25 Yamashina Yoshimaro, *Nihon no chōrui to sono seitai* (Tokyo: Azusa Shobō, 1934).

26 John James Audubon and William MacGillivray, *Ornithological Biography or an Account of the Habits of the Birds of the United States of America 5* (Edinburgh: A. Black, 1970), 327–8.

This is not to imply that Yamashina was a jingo. His pre-1945 writings were confined strictly to ornithological matters, and it has been argued that even his decision to choose ornithology over a career in the Imperial Japanese Army demonstrates his lack of interest in militarism. But even if Yamashina cared *only* about birds he nevertheless, like so many other ornithologists, took advantage of his polity's territorial and imperial expansion in order to secure specimens and construct taxa. Yamashina's ambit during this period was in fact circumscribed rather tightly by the Japanese Empire: he did not once travel outside its borders during the prewar period.[27]

Other WBSJ members were not aristocratic, but boasted elite intellectual pedigrees. One early member was the renowned ethnologist Yanagita Kunio, who took a keen interest in the role of birds in folklore as part of his broader project of surveying Japan's national folk culture. Another was the painter Araki Jippō, an exemplar of the 'birds-and-flowers' school that had been introduced to Japan from China in the Muromachi period (1336–1573).[28] But the majority of intellectuals who joined the WBSJ were scholars of English-language literature. The association's founder was Nakanishi Godō, a poet and lay Buddhist, who was by his own account heavily influenced by American Romantic poets such as Henry David Thoreau and Walt Whitman. Nakanishi had translated Whitman's *Leaves of Grass* into Japanese, and in 1929 had recreated his own Walden Pond experience, spending several months in seclusion in the countryside outside Tokyo.[29] A typical Nakanishi poem described 'the white-rumped swift on the unpeopled island, roosting hidden by white cliffs . . . the only human lamp coming from the fisherman's seasonal hut'.[30] Significantly, in Nakanishi's vast oeuvre of poems about birds, he wrote not only about places such as the Tōno Valley, where Yanagita had located the essence of the Japanese *volk*, but also about the frontier prefecture of Hokkaidō and the Okhotsk coastline.

It is true that the political implications of Nakanishi's State of Nature were subtly different to Thoreau's. Whereas Thoreau had compared

[27] YoshimaroYamashina, *Tori no heru kuni fueru kuni: Ōbei tori angya* (Tokyo: Nihon Chōrui Hogo Renmei, 1967), i.

[28] Nakanishi Godō, *Aichō jiden*, 468–480 cited in Higuchi, 'Birds for New Japan', 10.

[29] Tsutomu Takahashi, 'Thoreau's Confucianist Turn in Japan', *The Journal of Nineteenth-Century Americanists* 9 no. 2 (Fall 2021): 457.

[30] Nakanishi Godō, 'Kinkazan oki, Ashijima', in *Yachōki 16: shishū* (Tokyo: Shunjūsha, 1985), 151–2.

himself to a captive rooster, alerting the world to the joys of freedom, for Nakanishi the wilderness served to preach the virtues of communalism: he wanted to 'bring people to their homeland of nature, transcend a custom of private possession, and create a custom of public possession.'[31] But still, the activities of the WBSJ suggest that the 'nationalized nature' that began to emerge in interwar Japan owed no small debt to tropes of wilderness imported from Britain, Germany and the US.[32]

SCAP Promotes Nature and Democracy in Post-War Japan

In all these ways, then, we see the lineaments of a wildlife conservation movement forming in prewar Japan: a romantically infused aesthetic appreciation for Japanese landscape as 'wilderness', an interest in wild birds among a certain elite strata of the population and various state policies to classify and conserve the nation's natural heritage, including its rare or distinctive birdlife. But after Japan's defeat in World War II, these elements would be reconfigured in new and dramatic ways. A coalition of allied nations, led by the United States, occupied the country for an as yet unspecified duration. They then set about attempting to reform Japanese society root and branch, so as to purge it of its militaristic inclinations and transform it into a peaceful and democratic member of the community of nations. Supreme Commander Allied Powers (SCAP) assembled a bureaucracy that would oversee every aspect of this reform process – while also maintaining stability in a country that was still reeling from the shock of defeat in total war. This included setting up a Natural Resources Section and, within that, a Wildlife Branch led by one Oliver L. Austin, Jr.

Oliver L. Austin, Jr., was both a keen birder and – as a Harvard alumnus who summered at Cape Cod – a member of the closest thing American society has to aristocracy.[33] So it is perhaps not surprising that he got on extremely well with the community of ornithologists in Japan. Throughout his stint in Japan his family fraternised with theirs

31 Nakanishi Godō, *Yachō o tazunete* (Tokyo: Nisshin Shoin, 1942), 207; Nakanishi Godō, *Aichō jiden*, 477, cited in Higuchi, 'Birds for New Japan'.

32 For the concept of a 'nationalised nature' see Julia Adeney Thomas, *Reconfiguring Modernity: Concepts of Nature in Japanese Political Ideology* (Oakland: University of California Press, 2001), 30.

33 Mary H. Clench and J. William Hardy, 'In Memoriam: Oliver L. Austin Jr.', *The Auk* 106 no. 4 (October 1989): 706–9.

regularly, including frequent bird-hunting trips together (Figure 6.2). As well as Yamashina he also befriended Yamashina's uncle, the Marquis Kuroda Nagamichi, who was also an eminent ornithologist in his own right.[34] And he formed a particularly close working relationship with Kuroda's son, Nagahisa, who had spent the war training passenger pigeons for the Japanese Army.[35]

Austin had a theory, which he seems to have arrived at personally, that promoting nature conservation constituted an integral part of Japan's democratisation project. In a memo entitled 'Importance of Wildlife to the Mission of the Occupation' he argued that Japan's barbaric treatment of birds was a throwback to the era when wildlife had belonged exclusively to feudal lords, and was symptomatic of the country being mired in a militaristic, feudal stage of development. Modern democracy by contrast, he claimed, 'rejects any individual claim of ownership over natural things and considers birds and animals as nature's gifts for all people on an equal basis'.[36] After connecting wildlife protection with democratic values, he then went on to argue that the causal link ran both ways: promoting a love of wildlife in the hearts of Japanese youth was essential for instilling the spirit of democracy in the defeated nation. It would also help to prevent Occupation-era measures to conserve wildlife from being repealed once Japan regained its independence.[37]

[34] Annika A. Culver, 'Saving the Birds: Oliver L. Austin's Collaboration with Japanese Scientists in Revising Wildlife Policies in US-Occupied Japan, 1946–1950', *Endeavour* 41 no. 4 (2 November 2017): 156–7.

[35] Karasawa Kōichi, 'Kuroda Nagahisa', *Nihon Chōrui Gakkai shi* 61 Special Issue (June 2012): 17. For the history of wartime carrier pigeon training see Yoshida Kazuaki, *Sensō to denshobato, 1870–1945* (Tokyo: Shakai Hyōronsha, 2011).

[36] NARA-II RG331, Box 8981, Wildlife, vol. 1 (1946–1948); Oliver L. Austin, Jr., 'Importance of Wildlife to the Mission of the Occupation' (2 June 1947), cited in Higuchi, 'Birds for New Japan', 22.

[37] SCAP enacted a number of policies to conserve wild birds in Japan, including whittling down the list of permitted game species, and banning the use of nets (the preferred method of Japanese hunters, but seen as barbarically unsportsmanlike by Americans). SCAP's stringent restrictions on civilian firearm ownership also inadvertently helped to protect Japan's bird population. But Austin also told Uchida Seinosuke that 'if the reforms are to be permanent, they should be accomplished with as little SCAP interference as possible'. See Rinyachō, *Chōjū gyōsei no ayumi*, 326; Matsuyama Shirō, *Yachō to tomo ni 80 nen: kiitari mitari tameshitari* (Bun'ichi Sōgō Shuppan, 1997), 195; NARA-II RG331, Box 8981, Hunting, 'Regulation of Game Laws by American Forces 1946', 'Conference with Dr. Uchida 1948', 'Japanese Game Laws 1948', cited in Higuchi, 'Birds for New Japan', 24–5.

(a)

(b)

Figure 6.2a Oliver L. Austin Jr. during a hunting trip in Hokkaidō (c.1946–1950)
Figure 6.2b Prominent ornithologists and their wives gather at the home of Kuroda Nagahisa (c.1946–1950). Yamashina Yoshimaro is first from left; Kuroda Nagahisa is first from right. Oliver Austin and his wife are fourth and third from left, respectively
Source: Florida State University Oliver L. Austin Photographic Collection

With this in mind, Austin began working to establish civil-society institutions that might help to promote wildlife conservation. As early as November 1945, only three months after Japan's surrender, he invited a select group of ornithologists – all aristocrats – to a conference at the Shimbashi Daiichi Hotel, at the time requisitioned for use by SCAP high command.[38] Austin urged the participants not just to revive the Wild Bird Society of Japan, which had become moribund during the war, but also to establish a Japanese equivalent of the Audubon Society: an organisation dedicated not just to the appreciation of but also to the *conservation* of birds. Three months later, a conference sponsored by the Education Ministry announced the formation of a new society, the Japanese Society for the Preservation of Birds.[39]

With Austin's encouragement the JSPB membership, led by Yamashina, promptly set about lobbying to establish a National Bird Day in Japan. As we saw in Chapter 3, bird days had originally been established in several US states in the 1890s as part of a campaign to combat the moral malaise thought to be a product of the urban industrial environment.[40] By the 1920s, they had become a staple of school curricula not only in the metropolitan US but also in colonies such as the Philippines, where they helped to demonstrate America's 'higher humanitarian purpose in its empire'.[41]

In one sense the Japanese iteration of Bird Day represented a continuation of this colonial civilising mission. But it was innovative in that it explicitly connected an appreciation of Nature, manifested in the form of wild birds, with democratic values that the American occupiers perceived Japan to be lacking. The Education Ministry established Bird Day in April 1947, timed to coincide with the nesting season, and in 1950 this was expanded nationwide into Bird Week with the support of the Forestry Agency.[42] Schools across the country organised competitions for children to construct and install nesting boxes, the results to be judged by ornithologists. Toshihiro Higuchi

38 Matsuyama Shirō, *Yachō to tomo ni 80 nen*, 205. 39 *Ibid*, 307.

40 Doughty, *Feather Fashions and Bird Preservation*, 32.

41 Janet M. Davis, 'Bird Day' in Erika Marie Bsumek, David Kinkela and Mark Atwood Lawrence, *Nation-States and the Global Environment: New Approaches to International Environmental History* (New York: Oxford University Press, 2013).

42 Matsuyama Shirō, *Yachō to tomo ni 80 nen*, 307; Aoki Eiji, *Yamashina Yoshimaro no shōgai*, 216.

suggests that this activity may have been chosen in the hope that it would resonate with the public, many of whom had also lost their homes as a result of wartime aerial bombing. A SCAP Wildlife Section internal memo assessed nest boxes as 'of little practical value ecologically . . . Nevertheless the campaign deserves continuation, if only for its excellent propaganda value.'[43]

In a similar vein, Japanese birdwatching societies began opening junior clubs on the model of the American Audubon Society, the better to instil a reverence for nature (and therefore also democracy) in the hearts of the next generation.[44] In Hokkaidō, conservationists mobilised children to count endangered species such as the red-crowned crane, and schoolteachers made them write diaries projecting their own emotions on to the birds.[45] Indeed, some of the publicity material made the connection between nature and democracy fairly explicit: one book showed Yamashina helping elementary school children build a nesting box in the gardens of the National Diet Building (Figure 6.3).

Post-war planners even began to incorporate bird conservation into landscape architecture. Uehara Keiji, who since the 1920s had advocated the policy of 'nature for nature's sake' in national park planning, now called for Japan to import the German concept of the *Vogelschutzgehölze* (forest designed for protecting birds) as implemented in the Black Forest and upstate New York.[46] Bird conservation could be implemented on a small scale in cities, too, Uehara argued, by planting copses or even single trees in parks and gardens. Building on the research of Kuroda Nagahisa and Nakanishi Godō, Uehara suggested that urban planners could even choose which species they wanted to cultivate by selecting trees whose fruit they would find edible.[47] This vision were manifest in a new '*bādo sankshueri*' inside Tokyo's newly opened Yoyogi Park. The choice of site was highly significant: Yoyogi once been the site of a military parade ground; later, during the Occupation, it had been used for barracks housing SCAP forces.[48] The park thus symbolised Japan's path from militarism

[43] NARA-II College Park RG331, Box 8981, Wildlife, vol. 1, 1946–1948. 'Wildlife Conditions in Kyushu, Initial Report 18 April 1948' cited in Higuchi, 'Birds for New Japan', 23.

[44] Uehara Keiji, *Zōen taikei* (Tokyo: Kashima Shoten, 1974), 12.

[45] Janet Borland, 'Saving Red-Crowned Cranes: Children as Charismatic Conservationists in 1960s Japan', *Environmental History* 27 no. 1 (January 2022): 40–1.

[46] Uehara, *Zōen taikei*, 13. [47] *Ibid*, 29. [48] *Ibid*, 7, 18.

Figure 6.3 Yamashina Yoshimaro helps children install bird boxes in the gardens of the National Diet Building
Source: Aoki Eiji, *Ikite ita ahōdori* (1957)

to defeat to rebirth as peaceful, nature-loving democracy. And within the park, the sanctuary itself nestled against the wall of the neighbouring shrine to the Meiji emperor, physically connecting the imperial institution to these new post-war values. In recent years it has been colonized by a particularly aggressive strain of crows that, unsated by the fruits of its carefully curated trees, choose to eviscerate Tokyoites' garbage bags on a nightly basis.[49]

49 Aikawa Sadaharu and Fuse Rokurō, *Yoyogi Kōen* (Tokyo: Gōgakusha, 1981), 133. For a discussion of crows see Peter Wynn Kirby, *Troubled Natures: Waste, Environment, Japan* (Honolulu: University of Hawai'i Press, 2010), 97–9.

At the centre of all this was Yamashina Yoshimaro who, though stripped of his peerage by SCAP's constitutional reforms, rapidly carved out a new role for himself as Japan's most prominent campaigner for bird conservation.[50] Yamashina's interest in bird conservation did, to be fair, predate the US Occupation, but it was only after the war that he threw himself into the issue with full gusto. His reincarnation in many ways echoes the simultaneous effort by the Imperial Household Agency and the Occupation authorities to refurbish the image of Emperor Hirohito, transforming him from a militarist with pretensions to divinity to a more human figure – 'an avuncular biologist defined by his curiosity about the natural world'.[51] To this day, members of the Japanese royal family have continued to position themselves as patrons of nature conservation. At the time of writing Prince Akishino, brother of the current emperor, is currently serving as president of the Yamashina Institute.[52]

Part of the reason for Yamashina's post-war success can be attributed to the institutional and personal links he had forged with the American occupiers and, through them, with branches of Japanese state such as the Ministry of Education and Tokyo metropolitan government, that had previously shown little interest in wildlife protection. 'With fluent English and intercultural savoir faire,' writes Takie Sumie Lebra, many Japanese aristocrats 'survived the difficult times [after the end of the war] by filling the role of cultural interpreter and or liaison.'[53]

It may also be the case that many Japanese politicians and journalists had begun to internalise Austin's theory that nature conservation activities served to demonstrate commitment both to democratic values and to membership in the community of civilised nations more generally. After all, Japanese ornithologists had been lobbying for wild bird protection since long before the war, without succeeding in generating nearly so much of a reaction from the state or the mass media.[54] But in

50 Sugiyama-Lebra, *Above the Clouds*, 60.
51 Bix, *Hirohito*, 564; Ian Jared Miller, *The Nature of the Beasts: Empire and Exhibition at the Tokyo Imperial Zoo* (Oakland: University of California Press, 2011), 101.
52 Yamashina Institute of Ornithology, 'About Us': www.yamashina.or.jp/hp/english/about_us/president.html. Last accessed 16 January 2019.
53 Sugiyama-Lebra, *Above the Clouds*, 190.
54 An article suggesting a Japanese transplantation of Bird Day appeared in the very first issue of the ornithology journal *Tori* as early as 1915. See Watanabe Shōzarō, 'Tori no kinenbi' in *Tori* 1 no. 1 (1915): 3–6.

those early post-war years, many Japanese were acutely conscious of the Western colonising gaze. Even after the Occupation formally ended in 1952, the Japanese government remained preoccupied with how to 'normalise' its status within the international community. Paramount in many politicians' minds was the question of how to renegotiate the San Francisco Peace treaty, which still gave the United States sweeping powers to intervene in Japanese domestic politics in order to put down any 'riots and disturbances'.[55] There was also the question of how to reestablish diplomatic relations with Japan's former enemies and colonial possessions. Both might be accomplished more easily by showing that Japanese society had sloughed off those aberrant traits that had led to its aggression during the Pacific War.

Japanese ornithologists exploited this anxiety to argue for bird conservation, much as their Meiji predecessors had argued that the construction of a zoo would show that Japan belonged to the community of civilised nations.[56] Being able to demonstrate conservationist consciousness, Yamashina and his allies argued, would help to convince foreigners that Japan was fit to be trusted as a responsible member of the international community. Nakanishi Godō wrote that he wanted to transform Japan into a nation in which wild birds would no longer fear humans, so that foreign visitors to the country would 'consider Japan highly, become fond of the nation without reservations, and put a whole trust in it at once'.[57] Yamashina, who began traveling abroad widely after the war, constantly reiterated Austin's unfavourable comparison of Japanese attitudes to Nature with those of the West. In a book entitled *Countries where Birds Decrease and Countries where they Increase*, he castigated the Japanese for terrorising the birds of their homeland. During the Occupation, he wrote 'sparrows inside the metal wire surrounding the US base were not afraid of the soldiers. But outside the fence they returned to being normal sparrows, and fled from the enemy at the first glance.' (He did not mention that during the Occupation 'the enemy' was generally quite hungry.) Similarly, the lack of bird biodiversity in Tokyo's Shinjuku Park revealed the Japanese people's 'backwardness'. Whereas in the parks of London: 'you will see

[55] Nick Kapur, *Japan at the Crossroads: Conflict and Compromise after Anpo* (Cambridge, MA: Harvard University Press, 2018), 11.

[56] Miller, *The Nature of the Beasts,* Ch. 1.

[57] Nakanishi Godō, 'Yachō o ai suru kokoro', *Nyū Eiji* 1 no. 5 (n.d.) 56 cited in Higuchi, 'Birds for New Japan', 20.

wild ducks and squirrels unafraid of people. Children feed them with peanuts near Buckingham Palace, ducklings cross a road and a policeman stops traffic for them. People smile as they watch them. To have so many wild birds in the heart of city is the very definition of a civilised country.'[58]

Bird Conservation on the World Stage

Yamashina did not confine his bird conservation efforts to the domestic arena. He also sought to carve out a role for Japan in the burgeoning post-war international bird conservation movement. To this end he worked to place Japanese ornithologists at the centre of the movement's leading NGO, the International Council for Bird Protection (ICBP). In this way, Japan would not only show the world that it was adhering to civilisational standards; it would help to propagate those civilisational standards abroad, particularly in decolonized or decolonizing Asian nations. At the same time, international endorsement of Yamashina's efforts would help to further kindle enthusiasm for bird conservation back in Japan. The domestic and international campaigns would thus reinforce each other.

The ICBP (today Bird Life International) had been founded in London in 1922 as an umbrella organisation for the American Audubon Association, the Royal Society for the Protection of Birds, and bird protection societies in France and the Netherlands. In its early years it was heavily Anglo in complexion, with the remaining membership mainly drawn from Western Europe. The structure of the ICBP was both trans-national and national at the same time, in the sense that it was composed of national sections that were themselves composed of various member organisations in turn.[59] Japan's national section, for instance, was comprised of the JSPB, the Ornithological Society of Japan and Nakanishi Godō's WBSJ.[60] Other national sections comprised government or quasi-non-governmental organisations as well. The hybrid quality of the ICBP may

[58] Yamashina, *Tori no heru kuni fueru kuni*, 48–9, 219–20.

[59] International Council for Bird Preservation, 'Member Organisations', *Bulletin of the International Council for Bird Protection* I (London: ICBP, 1927), 5.

[60] International Committee for Bird Preservation, 'Proceedings of the XIIth Conference of the International Committee for Bird Preservation (Tokyo: 1960)', *Bulletin of the International Committee for Bird Preservation* VIII (London: ICBP, 1962): 31.

have contributed to its relative conservatism compared to later, more radical organisations such as Greenpeace or Friends of the Earth.[61]

Like the vast majority of its constituent institutions that campaigned to conserve wildlife, the ICBP had quite specific ideas about what sort of birds were worthy of protection, and from whom. Its founding principles expressed appreciation for the value of wild birdlife 'as incentives for reasonable field-sports', and framed the organisation's role as protecting the lives of animals as 'helping to preserve the balance between species which Nature is constantly seeking to adjust'.[62] For instance, it waged a protracted campaign against seagulls, which 'due to increasing adaptation to civilisation and human habitation ... [are] becoming more and more of a nuisance'. Concerned that the increase in seagull numbers was crowding out other, rarer species, it coordinated transnational campaigns to cull the birds.[63]

Japan had sent delegates to the ICBP from early on, represented at first by Prince Takatsukasa Nobusuke and then by Marquis Kuroda Nagamichi at the council's Geneva conference in 1928. But under Yamashina Yoshimaro's leadership the delegation took a much more active role in the organisation. In 1958 Yamashina took the lead in forming an Asian Section within the ICBP, initially with delegates from Japan, Korea, Malaya and India. (By 1961 Israel, Taiwan, Hong Kong and Thailand had also joined.)[64] The Japan national section began to subsidise the travel costs of delegates from some member countries who might otherwise not have been able to attend the biannual meetings.[65] With funding from Japan's Ministry of Agriculture and Forestry, Yamashina also launched an international effort to attach bird-bands to trans-border migratory species, a preliminary step to coordinating an international conservation strategy.[66]

61 Robert Boardman, *The International Politics of Bird Conservation: Biodiversity, Regionalism and Global Governance* (Cheltenham: Edward Elgar Publishing, 2006), 74.

62 International Council for Bird Preservation, 'Proceedings of the XIth Conference of the Bulletin of the International Committee for Bird Preservation (Helsinki, 1958)', *Bulletin of the International Committee for Bird Preservation* VIII (1962), 7.

63 ICBP, 'Proceedings of the XIth Conference', *Bulletin of the International Committee for Bird Preservation* VIII (Helsinki, 1962), 21.

64 ICBP, 'Proceedings of the XIIth Conference (Tokyo)', 42.

65 *Ibid*, 23.

66 *Ibid*, 117.

Yamashina also successfully lobbied for Japan to host the ICBP's annual congress in 1960. By a quirk of fate the delegates arrived in Tokyo just as the city was rocked by mass protests over the ratification of the US-Japan Security Treaty. Although the treaty represented a meaningful step in Japan's recovery of sovereignty after World War II, it committed the country to a lasting security alliance with the United States, placing it firmly on the front line of the Cold War in Asia. As many as one in three Japanese people participated in some form of protest against the treaty. [67] But while students and workers surrounded the National Diet protesting the terms of Japan's reintegration into the international community, the ornithologists toured the country in bucolic serenity. The Ministry of Agriculture and Forestry hosted a reception attended by Crown Prince Akihito and his new bride, after which the delegation visited a Zen garden in Kyoto, sampled a tea ceremony, and took a birdwatching tour in the chic mountain resort of Karuizawa, led by the poet Nakanishi.[68]

As the organiser of the conference, Yamashina used the podium to expound on the moral progress made by the Japanese people since the end of World War II. Though he lamented the Japanese government's 'deplorable' decision to cut back funding for wildlife conservation since the end of the Occupation, he was confident that civil society was nevertheless doing its part.[69] He cited one case in Hyōgo Prefecture, where wartime deforestation had driven Japanese white storks to start nesting on top of electricity pylons. Local residents had noticed the problem and, apparently of their own initiative, erected special nesting poles with some success. Moreover Japan, he argued, was now taking on the responsibility for promoting conservation overseas as well, for instance by having the Asian Section distribute posters drawn by Japanese schoolchildren to tour the other member countries.[70] The congress concluded by passing a resolution to lobby for Bird Day to be introduced on all national school curricula. Yamashina even urged the ICBP to lobby for a nature conservation clause to be inserted into the Antarctic International Treaty, which Japan had co-signed in 1959.[71]

[67] Kapur, *Japan at the Crossroads*, 1–2.
[68] ICBP, 'Proceedings of the XIIth Conference (Tokyo)', 31. [69] *Ibid*, 51–2.
[70] *Ibid*, 117. [71] *Ibid*, 35.

The Death and Life of Steller's Albatross

Certain species featured heavily in post-war Japanese conservation campaigns. The Oriental stork, the Japanese crested ibis and the red-crowned crane all attracted special attention, partly due to their auspicious associations within Japanese culture. It also helped that these birds tended to dwell in wetlands abutting rural communities, which made them suitable targets for encouraging grass-roots conservation campaigns.[72] But few species commanded as much attention from conservationists as Steller's albatross, a bird that before 1945 was virtually unknown to the Japanese public.

Yamashina's interest in these birds dated to before the war, for during a visit to Torishima in 1929 he had been able to witness a cull of the island's 2,000-odd remaining albatrosses first hand. In an article for *Tori* the following year he had written that: 'I could not bear to watch the butchering of these albatrosses. Only the word butchery can vividly express that scene .... I want to stamp out entrenched industries such as these from Japan as soon as possible.'[73] In the article he explicitly compared the 'butchery' of the Torishima settlers (who at the time of his visit numbered no more than a few dozen)[74] with that of the Japanese bird-hunters on Laysan Island as described by the conservationist William Hornaday – suggesting that even at this early stage he had internalised the American critique of Japanese people as particularly barbarous in their treatment of wildlife.[75]

Yamashina's lobbying did result in some limited efforts to protect the albatrosses of Torishima. Yamashina was able to draw on personal connections to have the Minister of Agriculture and Forestry to declare a ten-year hunting ban on the island, and soon after this the ministry banned albatross-hunting on the Bonins and Hachijōjima as well.[76] Shortly after this, the leaseholder of Torishima at the time promised not only to protect the islands' albatross chicks and eggs, but to install the necessary facilities to ensure their successful

72 *Ibid*, 51–2; Borland, 'Saving Red-Crowned Cranes', 36–7.

73 Yamashina Yoshimaro, 'Torishima kikō', *Tori* 31(1931): 8.

74 'Kimi yo shiru – minami no umi no paradaisu "Ahōdori no Torishima"'.

75 Yamashina, 'Torishima kikō', 8; William Temple Hornaday, *Hornaday's American Natural History* (New York: Charles Scribner's sons, 1927), 242.

76 Nōrinshō, 'Nōrinshō kokuji dai 290-gō' (4 August 1933); Yamashina, *Nihon no chōrui to sono seitai*, 1003, 1022; Hasegawa Hiroshi, personal correspondence with author.

nesting.[77] These early efforts were locally targeted, however, and received virtually no publicity beyond a few columns in Hachijōjima's local newspaper.[78] Steller's albatross remained on the central government's list of authorised game birds, and Torishima was not granted national monument status – even though thirteen other sites in Japan were listed on account of their seabirds during the interwar period.[79]

Moreover, it was far from obvious that those measures that were passed were even effective. The islanders got wind of the impending hunting ban and embarked on a concerted effort to kill as many albatrosses as possible before it was implemented. When Yamashina's assistant returned to Torishima in the spring of 1933, the local schoolteacher informed him that some 3,000 birds had been slaughtered over the course of the winter nesting season. (The other islanders denied all knowledge of this, claiming that a violent storm had somehow reduced the numbers of albatrosses on the island.)[80]

Torishima's albatross colony survived this final cull, for an article published in the *Hachijōjima Times* early in 1939 reported some 1,500 albatrosses as still living on the island.[81] But then the entire island erupted, triggering the wholesale evacuation of the human settlement on Torishima and smothering the albatross nesting site in lava.[82] Writing in *Tori* during the war, at a time when travel to the island was nigh-on impossible, Yamashina expressed his belief that this second eruption had:

> doubtless delivered the final, fatal blow to the albatrosses of Torishima, hovering as they did on the brink of extinction. [. . . .] Torishima seems to have been nearly the only breeding site for *Diomedea albatrus*. In which case I suspect this means that [this species of] albatross has not only ceased to nest on Torishima, but has befallen a truly tragic fate.[83]

77 Okuyama Shūsaku to Tokyo zeimu kantoku kyokuchō, 'Torishima shakutai ni tsuki chinjō' (24 December 1937) [Original housed in TMHB, copy provided to author by Hasegawa Hiroshi].

78 'Kimi yo shiru – minami no umi no paradaisu "Ahōdori no Torishima"', *Nankai Taimusu* (13 March 1939).

79 Yamashina, *Nihon no chorui to sono seitai,* 1046–47.

80 Yamashina Yoshimaro, 'Izu Shichitō no chōrui', *Tori* 11 no. 54 (December 1942), 244–5.

81 'Kimi yo shiru – minami no umi no paradaisu "Ahōdori no Torishima"', *Nankai Taimusu* (13 March 1939).

82 'Torishima kyū funkakō bakuhatsu! Hatsuka zen dekasegibito munen no hikiage', *Nankai Taimusu* (23 August 1939).

83 Yamashina Yoshimaro, 'Izu Shichitō no chōrui', 245; Oliver L. Austin, Jr., 'The Status of Steller's Albatross', *Pacific Science* 3 (1949): 288.

In 1949, four years after the end of the war, Oliver Austin Junior embarked on a tour of the Bonin Islands aboard a SCAP-supervised Japanese whaling expedition.[84] Though he, too, could not make landfall on Torishima due to rough weather, he wrote in the pages of *Pacific Science* that he 'was able to sail within close inspecting distance of every island in this chain on which Steller's Albatross is known to have bred . . . If Steller's Albatross were still breeding in this area, I should have seen some sign of it, either on the islands or the seas nearby . . . The chances that any of these fine birds remain alive today are remote indeed.' Not only did Austin's conclusion, albeit tentative, provide him with some kind of publishable result – an extinction is more attention-grabbing than simply a failed sighting, after all – it fitted nicely with his preexisting theory about Asiatic ecological rapacity:

> . . . the waters surrounding [the Bonin Islands] have swarmed with Japanese, Chinese and Okinawan fishermen for the last 50 years . . . it is hard to believe that any islet exists in this area which has not been visited many times by Oriental fishing boats in the 16 years since the last known Steller's Albatross was killed. It is equally unlikely that any of these craft would pass by an out-of-the way island with a bird on it without its crew attempting to land and kill the bird.[85]

Then, in the midst of the post-war campaign for bird protection, the Torishima albatross arose miraculously from the ashes. Barely a year after Austin had pronounced the species extinct, the crew of a weather observatory on Torishima noticed a few dozen of the birds nesting on a ledge beneath a sheer cliff. Three years later one of them, a keen birder, dispatched a carcass to the Ornithological Society of Japan in Tokyo for confirmation.[86] The rediscovery triggered a wave of what can be described, with

[84] 'SCAP Wildlife Expert Leaves for Torishima to Hunt Albatross', *Nippon Times* (1 April 1949).

[85] Austin, 'The Status of Steller's Albatross', 293–4.

[86] Shirai Kumihiko, 'Torishima no chōrui', in Jōtani Yukio and Shirai Kunihiko, *Torishima no shizen to ahōdori* (Tokyo: Tōkyō-to Kyōiku Iinkai & Asashi Shimbunsha, 1961), 8; Uchida Seinosuke, 'Horobiyuku ahodori', *Seiryō* 9 (March 1956), 13; Ōno Yoshiteru, 'Torishima no ',tori,' *Tori* 14 no. 66 (1955), 24–33; '"Extinct" bird still exists on Torishima', *Nippon Times*, 29 July 1954.

only mild exaggeration, as albatross-mania. Only a year after the end of the Occupation, the symbolism of the event was too good to pass up: here was a species that a member of the former occupying forces had declared vanished, an eternal victim of Japanese feudal barbarism, but which had been resurrected at the very moment Japan regained its independence.

There followed an outpouring of media interest in the albatross, which invariably stressed the importance of diligently protecting this rare species. The Kyōdō Film Company and NHK, Japan's public broadcasting corporation, both dispatched film units to Torishima to record wildlife documentaries about the bird.[87] Japan's national newspapers presented to their readership a heavily anthropomorphised portrait of the albatross that echoed the one crafted by American ornithologists half a century earlier (see Chapter 3). The Mainichi Shimbun compared the albatross' courtship ritual to the mambo, the global Cuban dance craze (Figure 6.4).[88] The *Yomiuri Shimbun* informed its readers that the albatross was not, in fact, a 'foolish' bird, but simply innocent of the wickedness of humankind. The bird's name had rather been coined by 'foolish men' who, 'killing hundreds of albatrosses each day, called their victims "fool birds" in order to rationalise their own wickedness'. The newspaper insisted that the very fact that the bird 'does not know to distrust people' meant that it was 'capable of great intimacy' with them.[89] The *Asahi Shimbun* was particularly effusive about the 'cheerful', 'friendly' albatrosses of Torishima, devoting numerous articles (including two front-page editorials) that emphasised both their intimate connection with humans and the urgent need to preserve Torishima as an 'albatross paradise'.[90]

Much of the credit for this wave of publicity goes to Yamashina, who as president of the JSPB spent most of 1955 touring Tokyo's media outlets and corridors of power drumming up interest in the bird. He petitioned both houses of the Diet, the Education Ministry, the Ministry of Agriculture and Forests, and the Forestry Agency, and

[87] 'Torishima no ahōdori (Kyōdō Eigasha)', *Asahi Shimbun*, 6 August 1956; 'Today's TV Choice', *Japan Times*, 27 January 1962.

[88] 'Ahōdori: manbo odotte "shūdan miai"', *Mainichi Shimbun* (13 November 1955).

[89] 'Ahōdori: hito natsukoi chikara mochi', *Yomiuri Shimbun* (29 July 1966).

[90] 'Aka-chan mo maruba genkina tori suna bi ahōdori', *Asahi Shimbun* (19 January 1961); 'Ahōdori no nakama (aichō shūkan ni yosete)', *Asahi Shimbun* (7 May 1961).

Figure 6.4 'The fools who dance and the fools who don't dance.' This cartoon puns on the Japanese word for albatross (*ahodori*, meaning 'fool-bird') and the lyrics to a popular dance performed at street festivals
Source: *Mainichi Shimbun* (13 November 1955)

appealed to the public through lectures, television and radio, newspapers and magazines.[91] He also arranged for publication of an anthology of children's stories, titled *The Albatross that Lived* (Figure 6.3). Like the newspaper coverage, the book's title story presented a highly anthropomorphised image of the albatross, describing its 'mambo-like' dance and likening its nesting behaviour to the cosy domestic rituals of a 1950s middle-class couple.[92] It also emphasised the international significance of the albatross' disappearance and rediscovery, recounting how Austin, 'the famous American bird researcher', had 'risked life and limb' [sic] scouring Torishima's steep cliffs before declaring that "the albatross has vanished from the face of the earth" . . . But then in 1954 a telegram travelled round the world . . . The albatross that was thought to have disappeared was alive!'[93]

As texts like these make clear, a key goal of post-war bird conservation campaigns was consciousness-raising, particularly among children

[91] Aoki, *Yamashina Yoshimaro no shōgai*, 231.
[92] Aoki Eiji, *Ikite ita ahōdori* (Tokyo: Aoba Shobō, 1957), 37. [93] *Ibid*, 34–5.

but also among the public more generally. In fact, many of the government policies they lobbied for also constituted a form of consciousness-raising, sometimes primarily so. Bird Week was a good case in point, but campaigns to have certain species demarcated as rare or precious also served a similar function. When, in 1956, the Tokyo metropolitan government provisionally designated Torishima's remaining albatrosses a natural monument, this also triggered another round of headlines.[94] The year after that, the Cultural Properties Preservation Committee created yet more headlines when it declared Steller's albatross and its nesting site on Torishima to be a 'thing for which Japan is renowned and thus in need of protection'. Five years later, in 1962, it upgraded its status to a '*special* natural monument'.[95] Finally, in 1965, the scope of the designation was expanded so that the entire island of Torishima was also declared a natural monument in its own right; moreover, albatrosses were henceforth to be considered protected wherever they were found in Japan.[96]

From 'National Treasure' to 'Endangered Species'

This campaign for ever more elaborate demarcation by the Japanese state dovetailed with a simultaneous campaign to demarcate Steller's albatross as a protected species at the international level as well. Through the auspices of the ICBP Yamashina also lobbied for the species to be added to the International Union for the Conservation of Nature (IUCN)'s nascent global biodiversity base. The IUCN had been founded by British biologist Thomas Huxley in 1948 with initial funding from UNESCO; as such, it enjoyed the imprimatur of intergovernmental support in a way that the ICBP did not. It also possessed

94 'Utamaro no haka nado 12-ken To no bunkazai ni shitei', *Yomiuri Shimbun* (24 February 1956); 'Ahōdori: manbo odotte "shūdan miai"'; 'Tensei jingo', *Asahi Shimbun*, 31 March 1962; 'Hito natsukoi kaichō', *Asahi Shimbun*, 20 March 1965; 'Tensei jingo', *Asahi Shimbun*, 27 March 1965.

95 'Ahōdori kakuage: shiseki tennen kinenbutsu o shin shitei', *Yomiuri Shimbun* (30 March 1962); *Nihon no tennen kinenbutsu* (Tokyo: Kodansha, 1995), 644; Fujisawa Tōru, *Ahōdori* (Tokyo: Tōkō Shoin, 1967), 32–3; Shōwa 33-nen bunkazai hogo iinkai kokuji dai 37-gō. While the Ministry of Education and Culture had been substantially dismantled during the Occupation, government oversight of cultural and natural monuments had been restored in the 1950 Cultural Properties Protection Law.

96 Izu Shotō Tōkyō Ikan 100-nen Shi Hensan Iinkai, *Izu Shotō Tōkyō ikan 100-nen shi*, 897.

a much wider mandate, aiming to protect not just wild birds but also the natural world more broadly. That said, it was in many ways substantively similar to the ICBP. Structurally it was also composed largely of national delegations, and in terms of both goals and membership the two organisations overlapped substantially. They regularly sent delegates to attend each others' conferences, and many of the naturalists who criss-crossed the international post-war scientific circuit were active in both simultaneously. Indeed, in practice the IUCN has historically delegated most matters pertaining to birds to its sister organisation – including biodiversity classification.[97]

In 1949, at its inaugural conference in upstate New York, the IUCN had drawn up a list of endangered species, the so-called Lake Success list. This list would eventually evolve into the Red List, a rolling effort to comprehensively list every threatened species on the planet. Efforts to catalogue biodiversity are not new, of course. In Europe they can be traced back to the works of Ray, Buffon, and most famously Linnaeus, and early modern Japanese scholars embarked on a similar project more or less independently.[98] But earlier generations of naturalists had proceeded in the serene assumption that they were cataloguing a static universe presided over by a benign Creator. Even Progressive-era bird conservationists who manifested vocal concern for the fate of wild birds were often surprisingly coy about extinction. Some found the theological implications of extinction blasphemous, and others viewed it as an inevitable side effect of civilisational progress. Just as many Social Darwinists assumed that Native Hawaiians were fated for extinction, so were evolutionary biologists generally fatalistic about biodiversity loss.[99]

In contrast, post-World War II efforts to catalogue biodiversity were propelled not only by concern about extinction but by the conviction

97 International Union for the Conservation of Nature Species Survival Commission, *IUCN Red List of Threatened Species* (Gland, Switzerland: IUCN, 2000), v.

98 Drayton, *Nature's Government*, 41; Federico Marcon, *The Knowledge of Nature and the Nature of Knowledge in Early Modern Japan* (Chicago: University of Chicago Press, 2015).

99 Alfred Russel Wallace wrote sardonically that 'Should civilized man ever reach these distant lands [in Indonesia], and bring moral, intellectual and physical light into the recesses of these virgin forests, we may be sure that he will so disturb the nicely balanced relations of organic and inorganic nature as to cause the disappearance, and finally the extinction, of these very beings whose wonderful structure and beauty he alone is fitted to appreciate and enjoy.' See Barrow, *Nature's Ghosts*, 71–2.

that something could be done about it. This mentality informed every aspect of the classificatory project from the beginning. Each entry on the list not only described each species' morphology, habitat and so on – it also estimated its population, reason for decline and breeding rate and listed protective measures either undertaken or proposed.[100] All entries were coded by level of precarity ('endangered', 'rare', 'depleted', 'indeterminate'), and indeed only later did the database expand to include non-threatened taxa ('out of danger' and finally, 'not at risk') as well.[101]

The list was also, from the outset, purposefully crafted to promote public engagement with nature conservation. Classifying biodiversity was partly an exercise in consciousness-raising, an attempt to galvanise the public and policy-makers to action. Aware that they could not possibly compile a comprehensive list of threatened species, IUCN delegates instead resolved to select particularly charismatic species whose plight might catch the imagination of the broader public.[102] Indeed, Ursula Heise describes the biodiversity database as 'a new variant of the modern epic or world text . . . the forever incomplete attempt to map the entirety of biological life and classify it according to its risk of extinction, as part and parcel of a battle of heroic scientists and conservationists against ignorant authorities and indifferent masses.'[103]

The project immediately caught the imagination of conservationists around the world. No sooner had the list been published than biologists and conservation NGOs began clamouring to have other species included in the Lake Success list. At the IUCN General Assembly in Caracas in 1952, for instance, another twenty-five species were placed on a provisional list for possible inclusion. It was therefore signally impressive that, in 1958, Yamashina succeeded in having Steller's albatross become the first new species added to the list since its inception nine years previously.[104] From thence, progress reports on Torishima's albatrosses would form a core component of future IUCN and ICBP meetings, reminding delegates that Japan was in the vanguard of the global wildlife conservation movement.

100 IUCN, 'Proceedings', *Eighth General Assembly (Nairobi, Sept 1963)* (1964): 71.
101 Warren B. King, Union for Conservation of Nature and Natural Resources. International Council for Bird Preservation; International and Service Commission. Survival, *Red Data Book. Vol. II: Aves* (Morges: IUCN, 1979), 2.
102 Boardman, *International Politics of Bird Conservation*, 52.
103 Ursula Heise, *Imagining Extinction: The Cultural Meanings of Endangered Species* (Chicago: University of Chicago Press, 2016), 65.
104 Boardman, *International Politics of Bird Conservation*, 58–9.

In this way, Yamashina's Occupation-era campaign to link wildlife conservation with post-war international reintegration continued, now waged on the national and transnational planes simultaneously. During Bird Week 1961 the *Yomiuri Shimbun* made the connection between wildlife conservation and diplomatic reintegration explicit, publishing a cartoon that portrayed albatrosses hectoring the Japanese people with incessant cries for 'normalisation' of the country's treaty relations with its neighbours (Figure 6.5).[105]

Just as nationalism and internationalism have long gone hand in hand, the one reinforcing the other, so have international organisations such as the ICBP and the IUCN reproduced the administrative structures of the nation state even as they seek to transcend them.[106] Constituted primarily of national sections, they can never be entirely aloof from national or nationalistic agendas. The tensions in this model became evident after, as we shall see in Chapter 7, another colony of Steller's albatrosses was discovered nesting on the disputed Senkaku (Ch: Diaoyu) island of Minami-kojima. At the 1980 conference of the ICBP Asian Section in Thailand, a motion was tabled to grant Minami-kojima 'special international reserve status' in order to better protect its albatross colony.[107] Yamashina's delegation retorted that Steller's albatross had been designated a natural monument of Japan, so granting one of its nesting grounds any kind of 'international' status was out of the question. Instead, the conference passed a watered-down motion calling simply for 'international cooperation, to make appropriate research and monitoring, as far as possible, of the colony in question'.[108] The dispute is arcane but does illustrate the way in which international conservation NGOs such as the ICBP are structured by the logic of national sovereignty. In this case, the Japanese delegation did not shy away from using its clout as a major international donor in order to assert cultural ownership of the species and, by extension, the islands it nests upon.

[105] The cartoon appeared next to an article on discussions to establish diplomatic relations between Japan and South Korea. For this reason the conical hats worn by the albatrosses reference stereotypical Japanese portrayals of Koreans. My thanks to Greg Pflugfelder for this observation.

[106] Mark Mazower, *Governing the World: The Rise and Fall of an Idea* (New York: Penguin, 2012), 48.

[107] Tickell, *Albatrosses,* 246.

[108] ICBP, 'Resolutions Adopted on 29 February 1980 at the IXth. Regional Conference of the Asian Section of the ICBP', *Asian Section News* 7 (November 1981): 13.

Figure 6.5 'Albatrosses – all year round the same chirping: "Normalization!"' This cartoon is a reference to ongoing diplomatic negotiations between Japan and the Republic of Korea, which coincided with the beginning of Bird Week. The triangular hats are an Edo-era trope used to represent Koreans
Source: *Yomiuri Shimbun* (5 May 1961)

Whither the Torishima Albatross?

So far I have suggested that the post-war campaign to conserve the albatrosses of Torishima was propelled not only by concern for the birds itself. The campaign owed as much to efforts by former aristocrats such as Yamashina Yoshimaro to reinvent themselves in the post-war era, and to concern about recovering Japan's sovereign status in international relations. A cynic might also observe that the

Japanese government's decision to designate Torishima as a nature reserve had an extraordinarily low opportunity cost. This is partly due to the island's own volatile tectonics, which have deterred any further attempts at permanent colonisation, but also because shifts in the global economy mean that few people are motivated to scale its cliffs any longer. Today's Hachijōjimans emigrate not southward into the Pacific, but northward towards mainland Japan. The albatrosses that once enticed Tamaoki's bands of bird-hunters are now too few to be harvestable on any kind of commercial scale, even if there were demand for them. And notwithstanding the Tokyo metropolitan government's stated concern about an imminent onslaught of tourists and mining companies, in the post-war period the island's natural resource value has generally been perceived as negligible.[109]

Moreover, for all the excitement about Steller's albatross, most people in Japan or elsewhere could only ever encounter the birds themselves in a highly mediated form – via books, newspaper cartoons, television documentaries and postage stamps. In the midst of the albatross boom, Torishima's only human population consisted of the weather station personnel. They were, in a sense, the most direct target of all the exhortations to care for the birds – this partly because they were suspected in some quarters of possibly tampering with the albatross colony.[110] After the observatory had opened in 1946, an alarming story circulated to the effect that 'the weather station personnel, noting [storm petrels'] tendency to fly to a bright light at night, tell how delicious meals of roast bird can be obtained simply by building a big bonfire after dark and letting the petrels fly into it of their own accord'.[111]

But it was also true that the weather station staff ended up performing almost all the actual physical labour involved in protecting the albatrosses for no financial compensation whatsoever. As on Laysan Island (see Chapter 4), this mostly involved attempting to eradicate animals

[109] In 1951, before the albatross conservation campaign got under way, a sulphur mine did operate briefly on the island. But it was abandoned after several months. See Kishōchō Torishima Kurabu 'Torishima' Henshū Iinkai, *Torishima*, 61.

[110] Tickell, *Albatrosses*, 246.

[111] Austin, 'The Status of Steller's Albatross', 289.

such as cats, rats and dogs that had been introduced to the island, and were deemed a threat to nesting chicks.[112] The staff also cooperated with a Yamashina Institute project to monitor the movements of the birds using bird rings provided by the US Department of the Interior.[113] For this they received no extra pay, but were instead garlanded in laurels from the ICBP and Japan's Agency for Cultural Affairs.[114]

In November 1965, however, smoke was noticed emerging from Torishima's caldera yet again, and as a precautionary measure the observatory staff were evacuated. Torishima and its albatrosses were cut off from the world once more. The island is now inaccessible except via helicopter or charter boat. Notwithstanding this, limited conservation efforts continue, organised by the Yamashina Institute and by ecologists such as Hasegawa Hiroshi of Tōhō University.[115] The focus has now shifted to protecting the albatrosses from the island of Torishima itself, constructing barriers to shield the nesting sites from avalanches and planting grasses in the ash slope to reduce erosion. Efforts to eradicate invasive fauna from Torishima still continue. And most ambitiously, a plan is under way to establish a new breeding site in the Bonin Islands in case the island itself erupts once again.[116]

It is not yet clear whether this will succeed. But regardless, an interesting inversion has occurred. Torishima had once been claimed for Japan through colonisation, both discursively through the rhetoric of development and in practice through the physical act of settling it. Now that the island has been abandoned its albatrosses continue, in life as in death, to knit the island into the fabric of the nation. William

112 Yoshikawa Motomu, 'Ahōdori no hogo' in Jōtani & Shirai, *Torishima no shizen to ahōdori*, 13.

113 Yamashina Yoshimaro, 'Chōrui hyōshiki shiken hōkoku dai 2-kai (Shōwa 37 nendo)', *Yamashina Chōrui Kenkyū* 3 no. 5 (December 1963): 310.

114 IUCN, 'Proceedings', 142–3; 'Observatory to Be Thanked by Group', *Japan Times*, 8 July 1962.

115 Since 2008 Professor Hasegawa, who has been an invaluable resource for the present study, has led an effort to reintroduce a population of Steller's albatrosses onto Mukojima in the Bonin Islands.

116 Deguchi Tomohiro, 'Zekkai e no saidōnyū – ahōdori', *Nihon no kishō chōrui o mamoru*, ed. Yamagiwa Satoshi (Tokyo: Tōkyō Daigaku Gakujutsu Shuppankai, 2009): 23–47.

Cronon writes that 'wilderness is the place where, symbolically at least, we try to withhold our power to dominate'.[117] By a similar token, the symbolic conservation of wilderness can also obscure a *failure* to dominate. On Torishima this may be of relatively slight consequence, for few people question the island's status as Japanese territory. But where sovereignty is contested, the stakes heighten considerably.

117 Cronon, 'The Trouble with Wilderness', 23.

7 *The Nature of the Senkaku Islands*

Biodiversity Conservation in Okinawa, 1945–2013

In the autumn of 2013 China Central Television aired a news bulletin with a provocative headline: 'Japan is snatching our islands using the pretext of environmental protection!'[1] This fiery denunciation was a reaction to an announcement by the Yamashina Institute that a new variant of Steller's albatross had been discovered nesting on a cluster of uninhabited islands in the East China Sea. For these were not just any islands. The Senkaku Islands (Ch: Diaoyu-dao or Diaoyu-tai, Eng: Pinnacle Islands, Ryukyuan: Yukun Kubashima or Igun Kubajima) are the subject of a long-running sovereignty dispute between Japan and China, a dispute that had flared up again three years previously when the Japan Coast Guard had arrested a Chinese fishing captain in the waters around the islands. Ambassadors had been summoned, nationalist demonstrators had taken to the streets in Tokyo and Beijing, and fighter jets had been scrambled in the airspace over the disputed territory. Yet the Yamashina Institute's director had glossed over all this when he called for Japanese ecologists to conduct an on-the-ground survey of the islands. 'Quite apart from political factors,' he told a Japanese newspaper reporter, 'the birds must be protected.'[2]

What are we to make of such conflicting claims? Was the Yamashina Institute engaged in a high-minded conservationist endeavour that transcended diplomatic pettifogging? Or was it a devious ploy – a 'pretext' for a Japanese land-grab? It is tempting to dismiss hyperbolic claims by an official media outlet of the People's Republic of China as mere propaganda. Yet the Yamashina director's statement – to the effect that efforts to conserve nature on the islands can somehow be disentangled from the politics of sovereignty – also deserves

1 CCTV 4 Zhongwen Guoji, 'Ri yanjiu xiaozu na Diaoyu-dao 'Xintianweng houdai' zuo wenzhang' (15 September 2013): http://news.ifeng.com/mil/video/detail_2013_09/15/29627973_0.shtml. Last accessed 3 November 2014.

2 'Haha ahōdori wa Senkaku umare, Ogasawara Mukojima' *Nihon Keizai Shimbun* (14 September 2013).

interrogation. We have already seen that the relationship between nature conservation and sovereignty has, historically, been a complex one. It continues to be so to this day.

This chapter picks up where Chapter 6 left off, at the peak of albatross mania in post-war Japan. But it shifts the focus from Japan's North Pacific borderland to its still very much contested border in the East China Sea. In 1963 the Central Meteorological Office in Tokyo despatched a letter to its counterpart on the Okinawan island of Ishigaki enquiring as to whether any specimens of Steller's albatross were to be found nesting on another group of abandoned bird islands – the Senkaku Islands. By the time the letter was sent mainland Japan had in many ways succeeded in normalising its international relations, and in recovering a large portion of the sovereignty lost at the end of World War II. Though it was still very much the junior partner in a security pact with the United States, the treaty of 1960 had rolled back the US prerogative to intervene in domestic politics, and the number of US troops stationed on the mainland had been drawn down substantially.[3] Japan had also established diplomatic relations with its former colony of Taiwan, and was about to do the same with South Korea. And in possibly the most prestigious mark of the former aggressor's peacetime rehabilitation, Tokyo was on track to host the Olympics the following year.

By sharp contrast, however, Okinawa remained in a state of sovereign limbo. At one point the archipelago had been ruled by the Ryūkyū Kingdom that, while paying tribute to China, enjoyed a high degree of autonomy and traded widely across East and Southeast Asia as far as Java.[4] But in the seventeenth century the islands became a de facto colony of the Japanese domain of Satsuma, and in 1879 the kingdom was abolished and the archipelago renamed Okinawa Prefecture as part of the Meiji government's general effort to align its territorial borders with emergent principles of international law.[5] Over the

[3] Kapur, *Japan at the Crossroads*, 17.

[4] Gregory Smits, *Maritime Ryukyu, 1050–1650* (Honolulu: University of Hawai'i Press, 2019).

[5] Mamoru Akamine, Lina Terrell and Robert N. Huey, *The Ryukyu Kingdom: Cornerstone of East Asia* (Honolulu: University of Hawai'i Press, 2019), 148–9. The extent of Ryūkyūan autonomy in the early modern period is a hotly debated and politically loaded topic. For a recent overview, see Travis Seifman, 'Performing "Lūchū": Identity Performance and Foreign Relations in Early Modern Japan' (PhD diss., University of California Santa Barbara, June 2019): 21–4.

decades that followed Okinawans endured mainland chauvinism, forced cultural assimilation and economic marginalisation – capped by an appalling degree of wartime suffering that rivalled Iwo Jima and Saipan.[6] In the last desperate days of the war, the Japanese military high command mobilised the entire population of Okinawa Main Island to resist the US invasion, hoping to inflict enough casualties to deter the enemy from launching a ground invasion of the mainland. During the battle that followed much of the island was razed and as many as a third of the civilian population perished.[7]

In the wake of the battle US diplomats puzzled over what exactly to do with Okinawa. Early on a decision had been taken to administer the archipelago separately from mainland Japan, but wartime plans to restore the islands to independence, cede them to China or even roll them into a newly established fifty-first state of the Union never materialised.[8] As the Cold War in Asia heated up, the archipelago started to take on increasing strategic importance in the eyes of US military planners. The islands came to host the bulk of US military forces in the Pacific and to be a front line in the struggle to contain communism in Asia. The United States Civilian Administration of the Ryukyus (USCAR) was established, and the US began preparing to occupy the islands indefinitely.[9] Okinawa would eventually be returned to Japan in 1972 but the exact date of the handover remained uncertain until the late 1960s, even as the US military footprint expanded as a result of the Vietnam War.[10]

Okinawan Culture Wars

American diplomats were vexed as to how to rationalise this prolonged military occupation. As USCAR dug in for the duration it began casting around for ways to increase the legitimacy of its

[6] Alan S. Christy, 'The Making of Imperial Subjects in Okinawa', *Positions* 1 no. 3 (Winter 1993): 611–12.

[7] Tze May Loo, *Heritage Politics: Shuri Castle and Okinawa's Incorporation into Modern Japan, 1879–2000* (Lanham: Lexington Books, 2014), 149.

[8] Hara, 'Okinawa, Taiwan, and the Senkaku/Diaoyu Islands', 4; Immerwahr, *How to Hide an Empire*, 228.

[9] Nicholas Evan Sarantakes, *Keystone: The American Occupation of Okinawa and US–Japanese Relations* (College Station: Texas A&M University Press, 2001), 59.

[10] Thomas R. H. Havens, *Fire Across the Sea: The Vietnam War and Japan 1965–1975* (Princeton: Princeton University Press, 2014), 87–8.

administration, while also helping Okinawans to accept prolonged separation from mainland Japan. One strategy involved partial democratisation, allowing Okinawans to elect their own governor and independent assemblies.[11] But this approach had its limitations. Internal USCAR memos assessed that more than three quarters of the population desired reunification with mainland Japan, 'basically motivated by nationalistic feelings', with only 9 per cent desiring continued American rule, US statehood or independence.[12] Electoral campaigns therefore became a platform for political parties to protest the ever-growing US military presence and demand reversion to Japan.[13] Indeed, democratisation set in motion a chain of events that did indeed lead to reversion in 1972 – although not to an appreciable reduction in the military footprint on the islands.

Another answer that USCAR hit on to boost its legitimacy was cultural policy, in the hope that cultivating a sense of distinct Ryukyuan identity might blunt Japanese nationalist sentiment among Okinawans.[14] In 1949 it established a Department of Culture to promote Ryukyuan heritage, and the Communications Ministry ran a competition to solicit new postage stamp designs to replace the old Japanese ones. The winning entries featured stereotypically 'Ryukyuan' artefacts such as karate, terracotta-tiled roofs, dragons' heads, 'Chinese' lions and the requisite local women wearing traditional dress (Figure 7.1).[15] In the same year USCAR partnered with local intellectuals to form a Committee for Conserving Historical Relics. The next year it established the University of the Ryukyus with help from experts dispatched from Michigan State University.[16] The new university was pointedly sited atop the ruins of Shuri Castle, the former seat of Ryukyuan kings that had been razed to the ground during the American invasion.

[11] David Tobaru Obermiller, 'The US Military Occupation of Okinawa: Politicizing and Contesting Okinawan Identity 1945–1955', (PhD diss., University of Iowa, May 2006): 265.

[12] OPA 6-004-211271a036: 'Okinawa Reversion: An Analysis of Japanese and Okinawan 'Views': 1, 13'.

[13] Obermiller, 'The US Military Occupation of Okinawa', 33, 285, 301.

[14] Loo, *Heritage Politics*, 151.

[15] Ryūkyū Seifu Bunkazai Hogo Iinkai, *Bunkazai yōran 1965-nenban* (Naha: Ryūkyū Seifu Bunkazai Hogo Iinkai, 1966), 2.

[16] Ogawa Tadashi, *Sengo Beikoku no Okinawa bunka senryaku: Ryūkyū Daigaku to Michigan misshon* (Tokyo: Iwanami Shoten, 2012), 66.

Figure 7.1 Stamps issued to commemorate distinctly 'Ryukyuan' cultural and natural heritage

Because USCAR wanted to give the impression of promoting Ryukyuan self-rule, it chose to delegate the details of this cultural policy to Okinawans. In 1952 a Ryukyu Cultural Properties Protection Committee (RCPPC) was established, staffed by local intellectuals: academics, museum curators, journalists and the like.[17] Finding suitably qualified individuals was not an easy task, however. USCAR turned to the academic and former diplomat George Kerr as a consultant on cultural policy, and in confidential memos Kerr chafed against what he saw as the incompetence, cronyism and even

[17] Loo, *Heritage Politics*, 2.

corruption of those Okinawans who did end up staffing the Cultural Properties Protection Committee.[18] He observed tartly that: 'there are sometimes signs in the local (Ryukyuan) government of a desire to reserve the area to local "scholars". There are none ... Recent Japanese projects have been summer holiday affairs, prosecuted, I fear, with more attention to good newspaper coverage than to technical accuracy in fieldwork.'[19] In some instances, Kerr complained, local secondary-school boys were even despatched to conduct their own investigations of archaeological sites.[20]

Matters came to a head during the construction of the Shuri Castle Museum, when one of Kerr's more withering memos was leaked to the Okinawan press, engulfing the entire project in scandal.[21] Kerr castigated the RCPPC's leadership for 'indulging in personal hobby interests', and urged that an American colleague from the Bishop Museum in Honolulu be brought in to oversee the plans for the new museum.[22] The US Army High Commissioner likewise expressed a preference for an American to oversee any cultural activities funded by the government of the Ryukyus, '[f]or a number of reasons, cultural, technical and political'.[23]

The politics of expertise also mingled with the politics of identity. While USCAR hoped to promote distinctively Ryukyuan culture, many of the Okinawan committee members tended to emphasise the archipelago's cultural and historical links with mainland Japan. Annotations of (Japanese-language) draft committee reports show terms like 'Okinawa' and 'Our Country Japan' slashed out in red, replaced by 'Ryūkyū' instead. References to certain artefacts having been designated as 'national treasures' in the prewar period are also excised.[24] In 1963 Kerr complained that a large proportion of designated heritage so far comprised of Yaeyama shell mounds, the particular hobby-horse of the committee chair but also crucial evidence in an

18 OPA G .H. Kerr (Asia Foundation): Kerr to Oshire (16 July 1962); Kerr to Binford (9 October 1963); Kerr to Ripley (10 August 1964).
19 OPA G. H. Kerr (Asia Foundation): Kerr to Force (5 September 1963).
20 OPA G. H. Kerr (Asia Foundation): George H. Kerr, 'Memorandum: Cultural Assets Conservation Work in the Ryukyus – Problems, Programs, and a Reorganization Proposal' (16 July 1963), 5.
21 'Senmon gyōmu no kōjō o: Jōji Kea hakushi bunho gyōsei kibishiku hihan', *Okinawa Taimusu* (13 October 1963).
22 OPA G. H. Kerr (Asia Foundation): Kerr to Stewart (30 July 1963).
23 OPA G. H. Kerr (Asia Foundation): Caraway to Kerr (18 June 1963).
24 OPA J 12 8 007–06 Bunka Hogo Iinkai, 'Kōhō kengi 1955–1963'.

archaeological theory that linked the entire archipelago to the Jōmon prehistoric culture of mainland Japan. Kerr, who also happened to be a specialist on Taiwanese history, complained that artefacts linking the Ryūkyū islands with cultures to the south were being neglected.[25] He also criticised a tendency to prioritise wilderness heritage and prehistoric cultural assets, which he believed gave the impression of Okinawa as an essentially primitive culture rather than a thriving autonomous one.[26][27]

Matters became more complicated as reversion to the mainland became imminent. From 1967 the RCPPC pointedly began classifying heritage as Okinawan rather than Ryukyuan.[28] At the same time, mainland Japanese schools began planning Okinawa-themed festivals and cultural activities, aimed at drumming up interest in the soon-to-be-returned prefecture.[29] Japan's Agency for Cultural Affairs also began joint-funding heritage conservation projects, and was naturally keen to emphasise those sites that showed links to the mainland. Anticipating eventual amalgamation, the RCPPC began holding joint meetings to liaise with its Japanese counterpart. Kerr worried that 'those Japanese who have worked in the Ryukyus seem determined to only search prehistoric sites for evidence of ancient ties with Japan'. He again urged USCAR to solicit help from American research institutions, even while worrying that the RCPPC would flat out refuse to collaborate with non-Japanese scholars.[30]

Woodpecker Protection and Anti-Base Activism

If these struggles over how to define Okinawa's cultural heritage were waged primarily at the level of discourse, the stakes became substantially higher when nature became the object of conservation. Indeed,

25 OPA G. H. Kerr (Asia Foundation): Kerr to Ripley (10 August 1964).
26 OPA G. H. Kerr (Asia Foundation): George H. Kerr, 'Memorandum: Cultural Assets Conservation Work in the Ryukyus – Problems, Programs and a Reorganization Proposal', 4.
27 For a similar dynamic that played out in fieldwork conducted on the border island of Tsushima, 1950–1951, see Ankei, 'Fieldwork Ethics and Positionality of Tsuneichi Miyamoto', 34.
28 Ryūkyū Seifu Bunkazai Hogo Iinkai, *Okinawa no bunkazai* (Naha: Keitai Hyōronsha, 1967).
29 OPA G. H. Kerr (Asia Foundation): USCAR Okinawa Press Highlights (4 November 1963).
30 OPA G. H. Kerr (Asia Foundation): Kerr to Warner (11 October 1965).

nature conservation campaigns could also have a direct political impact on events on the ground in Okinawa. One particular species, the Okinawan woodpecker (*Dendrocopos noguchii*; J: *noguchigera*), became a wedge issue around which anti-base activists could organise to protest the US military presence.

The Okinawan woodpecker inhabited a highly circumscribed range in the Yanbaru Forest in the municipality of Kunigami, in the northern part of Okinawa Main Island. In 1955 it became one of the first species to be designated a Ryukyu natural monument, although the burst of publicity was not accompanied by much in the way of on-the-ground conservation measures.[31]

Then, in 1957, the US military requisitioned roughly half of the woodpecker's habitat for use as a Marine Corps training base.[32] Naturalists soon began raising questions about the potential ecological impact on the woodpecker, and as in Chapter 6 they readily exploited transnational conservationist networks in order to bolster their cause. In 1964 the Asian Section of the ICBP, still led by Yamashina Yoshimaro, met in Hong Kong and passed a resolution that cited 'recent studies by Okinawan biologists' to warn USCAR of 'the importance of strict protection of the Noguchii woodpecker' against any use of its habitat for military purposes. The following year George Ruhl, representing both the US National Park Service and the IUCN, attended the ICBP World Congress in England and reported on a successful sighting of Okinawa woodpeckers – adding also that the species 'was attracting international attention, particularly from photographers on the Japanese mainland'.[33]

Throughout the 1960s the US military made little further attempt to use more of the land it had seized, even going so far as to tacitly permit local farmers to continue cultivating their landholdings as before. But in December 1970 the Marine Corps announced plans to open a firing range in the Yanbaru Forest that could train soldiers for jungle combat

[31] 'Kokuhō jūbun tennen kinenbutsu: maboroshi no yachō noguchigera, Okinawa de san wa hakken' *Mainichi Shimbun* (26 June 1967); Lester L. Short, 'Habits, Relationship and Conservation of the Okinawan Woodpecker', *The Wilson Bulletin* 85 no. 1 (March 1973): 5–20.

[32] Gavan McCormack and Satoko Oka Norimatsu, *Resistant Islands: Okinawa Confronts Japan and the United States* (Lanham, MD: Rowman & Littlefield, 2018), 79.

[33] Cambridge University Library, M.S. Peter Scott papers: 'ICBP Report of XIVth International Conference, Cambridge, England 11–15 July 1966', 27–8, 30.

in Vietnam.[34] The announcement triggered immediate protests from local residents, many of whom faced the prospect of being summarily evicted from land they had cultivated for generations.

Anti-base activists were quick to enlist woodpeckers, and through them international conservation organisations, in support of their cause. One Kunigami resident happened to be a member of the Japan Wild Bird Association and wrote to Nakanishi Godō in Tokyo, who promptly fired off letters of protest to USCAR and the White House, as well as to the IUCN in Switzerland.[35] The mayor of Kunigami argued that firing rounds in the Yanbaru Forest would 'frighten to death' the woodpeckers, and insisted that his municipality 'was bound by advice . . . from the international society in Switzerland [probably the IUCN or perhaps the ICBP] that they should do everything to preserve the Noguchigera woodpecker'.[36] And on mainland Japan the *Mainichi Shimbun* mourned the fact that great numbers of woodpeckers had 'died in battle' during the Pacific War; surely an American firing range would finish the last few off entirely.[37]

In the face of this opposition, the Marine Corps reluctantly decided to postpone plans to conduct live firing indefinitely – though it did not rule out the use of blank firing and helicopters.[38] Opposition to the firing range continued to seethe even after reversion in May 1972. Ten days after reversion Japan's Ministry of Defence put out a statement to the effect that Japanese Self-Defense Forces would not participate in joint exercises in the Yanbaru Forest.[39] Meanwhile the Culture Agency began planning to buy up the remaining eighty hectares of the forest as a nature reserve in an effort to conserve the woodpecker's habitat.[40]

The woodpecker conservation campaign in many ways foreshadowed later efforts by anti-base activists to use ecological damage

34 'Mokunin kōsakuchi toriage: Okinawa Beigun tsūkoku 'jitsudan enshū ni tsukau' *Yomiuri Shimbun* (23 December 1970).
35 '"Noguchigera" o mamotte', *Mainichi Shimbun* (30 December 1970). The newspaper's reportage added with a flourish that a great numbers of woodpeckers 'died in battle' during the Pacific War, with the remaining few facing extinction should the American firing range be established.
36 NARA-II RG 260 187 '"Noguchigera" Woodpeckers: Visit to Kunigami Training Area with MG Wilson' (1 February 1971).
37 '"Noguchigera" o mamotte', *Mainichi Shimbun* (30 December 1970).
38 NARA-II RG 260 187 "'Noguchigera" Woodpeckers: Visit to Kunigami Training Area with MG Wilson' (1 February 1971).
39 'Noguchigera mamoru', *Mainichi Shimbun* (25 May 1972).
40 McCormack and Norimatsu, *Resistant Islands*, 79.

as a wedge issue to protest the US military footprint in Okinawa. As recently as 2019 activists successfully lobbied the IUCN to classify Okinawa's dugong as 'critically endangered' in order to oppose the relocation of the Futenma Airbase to Henoko Bay.[41] Woodpecker numbers, however, continue to dwindle to this day, although it is unclear what role the US military presence has played in this decline.[42] The requisition of a large slice of Yanbaru as a military training ground may actually have protected the original primary forest canopy, which outside the reserve was largely felled and replaced with Japanese cedar.[43]

The Post-War East Asian Context of the Senkaku/Diaoyu Islands Dispute

So by the time the Tokyo Meteorological Observatory began asking about albatrosses on the Senkaku Islands, the politics of nature conservation in Okinawa were already febrile. But by enquiring about the situation on the Senkaku Islands in particular, the letter injected yet more controversy – because if the sovereign status of Okinawa was ambiguous, the status of the Senkaku Islands was (and remains) even more so. The islands lie roughly equidistant from Taiwan, nominally a part of China, and the Yaeyama Islands, the southwesternmost island group in the Japanese archipelago. The Japanese government had first claimed them in 1895, annexing them to Okinawa Prefecture during the First Sino-Japanese War (1894–1895). But the Qing government never recognised the annexation, and its successor governments in Beijing and Taipei both maintain that the wartime context renders the Japanese claim illegitimate. Since World War II the US government has also trodden carefully around the issue, reluctant to get drawn into the dispute. Although the State Department has treated the islands as 'administratively' part of the Ryukyus/Okinawa, since 1971 it has

41 'With a population under a dozen, Okinawa dugongs labeled "critically endangered" by IUCN', *Japan Times* (12 December 2019).

42 IUCN Red List, 'Okinawa Woodpecker': www.iucnredlist.org/species/22681531/125513230. Last accessed 1 July 2020.

43 Ogasawara Ko and Ikehara Sadao, 'Ecological and Behavioral Observations of Okinawa Woodpecker *Sapheopipo noguchii*, with Notes on Conservation', *Yamashina Chōrui Kenkyū* 9 no. 2 (1977): 24; 'Maboroshi no tori noguchigera aware – shinrin ranbatsu ni oware – nokoru 'tenkoku' Beigun kichi', *Okinawa Taimusu* (10 November 1973).

adopted a policy of taking 'no position on sovereignty' so as not to inflame its relations with either the Chinese or Taiwanese government.[44]

Like other bird islands discussed so far, the Senkaku Islands were first targeted for their resources. During the 1880s Japanese fishermen began visiting the islands to collect turban snails that could be used to manufacture mother-of-pearl. After the turban snail population was depleted they switched to albatrosses, decimating that population, too, before moving on to other seabirds that could be stuffed and sold as ornaments. At its peak more than 240 people lived on Uotsuri-shima (Ch: Diaoyu-dao), the largest island. But after 1910 the settlement was all but abandoned, its profitability undermined by the dwindling bird population and a collapse in demand for avian apparel.[45]

Still, the islands continued to be a fixture of the economic geography of the East China Sea. A small bonito processing plant operated on Uotsuri-shima until 1940, and fishing vessels from Ishigaki and Taiwan plied the waters around the islands, their crews sometimes making landfall to stock up on provisions. Desperate food shortages that afflicted Ishigaki towards the end of World War II drove some islanders to make forays to the islands to harvest seabirds and eggs.[46] After 1945 the border between Taiwan and Okinawa was closed, making the islands an even more attractive destination. Fishermen from newly liberated Taiwan and newly occupied Okinawa frequently met up on the island to trade pleasantries, information and smuggled goods: Taiwanese rice, sugar and tea in exchange for black-market US army surplus.[47]

Though the Senkaku Islands were well known along the East China Sea littoral, in the immediate post-war years they attracted little attention from governments. What little interest there was in the islands was due largely to the efforts of one man. Takara Tetsuo (1913–2014) was one of post-war Okinawa's most prominent academics and public intellectuals. Over the course of a long and accomplished career, he developed a mind-bogglingly broad range of interests, but throughout

44 Hara, 'Okinawa, Taiwan, and the Senkaku/Diaoyu Islands', 11.

45 Hiraoka Akitoshi, 'Meiji-ki ni okeru Senkaku Shotō e no Nihonjin no shinshutsu to Koga Tatsushirō', *Jinbun Chiri* 57 no. 5 (2005): 514.

46 Iramina Takakichi, 'Chichi no ikotsu shūshū wa, jibun no fune to te de', in Senkaku Shotō Bunken Shiryō Kenkyū Hensankai, *Senkaku kenkyū: Takara gakujutsu chōsadan shiryōshū vol. I* (Naha: Dētamu Rekiosu, 2007), 361.

47 Nishimura Kazuyuki, 'Taiwan higashi kaigan ni okeru gyorō gijutsu no bunka shigenka', *Higashi Ajia Kindai Shi* 17 no. 3 (2014): 48.

Figure 7.2 Takara Tetsuo by the ruins of Koga Tatsushiro's bonito processing plant on Uotsuri-shima (1950)
Source: *Senkaku kenkyū* 1 (2007)

his life he repeatedly returned to an early interest of his – exploring the economic and ecological value of the Senkaku Islands (Figure 7.2).

Takara Tetsuo was born on Ishigaki, the most populous island of the Yaeyama group as well as its administrative centre. He received his early education on the island but then transferred to the prefectural agronomy school on Okinawa Main Island. In 1934 he moved to mainland Japan, enrolling in higher school in Kagoshima, where he specialised in agricultural biology. In 1938, a year after graduating, he entered military service at the age of twenty-three.[48] After the war he returned to Ishigaki to take up dual posts as the headmaster of the local agronomy college, and as extension officer for USCAR's Ministry of Agriculture. In this latter capacity he turned his efforts to boosting the

[48] OPA J 12 2 001–12, Bunkazai hogo iinkai iin shokuin rirekisho-shū: 'Takara Tetsuo rirekisho'.

local economy, a particularly pressing concern given that the war had left many islanders on the brink of starvation.[49] A series of articles he penned for the local journal, *Yaeyama Culture,* were eventually compiled into a monograph on agricultural development, aimed not only at locals but also at prospective immigrants to the islands.[50]

This kind of scientific economic boosterism consciously echoed the work done by the soil scientist Tsunetō Noritaka, who forty years earlier had toured Okinawa in search of untapped phosphate reserves. But Takara also began branching out into avenues of research that had, at best, a tangential connection to economic development. It was in the pages of *Yaeyama Culture* that he penned the first of many articles about the habu, a poisonous snake highly sought after as an ingredient in a popular Okinawan liquor.[51] And he would have almost certainly come to know the head of Ishigaki's meteorological observatory, a man named Senanami Chōsen, who beyond his regular duties also maintained a keen interest in birds. Around the same time that Takara was writing for *Yaeyama Culture* Senanami penned an article urging readers to observe the correct hunting seasons and refrain from bagging protected species, partly because birds were beneficial for agriculture and fishing (they helped to identify where shoals of fish were located) but also because they were worthy of conservation in and of themselves.[52]

In other words, Takara was part of an intellectual community interested not only in optimising the exploitation of natural resources for human benefit, but also alive to the importance of protecting nature in its own right: if there was a little bit of Tsunetō Noritaka in him, there was a little bit of Yamashina Yoshimaro in there also. This is evident in the explanation he gave for embarking, in 1950, on the first of what would eventually become five expeditions to the Senkaku Islands. While his primary aim was to investigate the islands' economic potential as 'undeveloped land' for an Okinawa 'scorched and starved by war', he also planned to survey the bird and marine life of the islands. There was also an element of sheer curiosity. It had, he later claimed, been a boyhood dream of his to visit the islands, and he also hoped that

49 'Jinmin sensen no hata', *Yaeyama Bunka* 1 (July 1946).

50 Takara Tetsuo, 'Kyōdō nōgyō no kagaku gijutsuka', *Yaeyama Bunka* 10 (April 1947); Takara Tetsuo, *Yaeyama kaihatsu shiryō* (n.d.).

51 Takara Tetsuo, 'Habu no bunpu zu', *Yaeyama Bunka* 32 (June 1949).

52 Senanami Chōsen, 'Gunchō zassō', *Yaeyama Bunka* 11(May–June 1947): 7.

his ornithological observations might be of some use in reinvigorating the post-war education of young boys.[53]

This early sally was a modest affair – Takara simply hitched a lift from Ishigaki aboard a fishing vessel headed for the largest island of Uotsuri-shima. As he was unable to visit the other islands in the group his survey of the islands was hardly comprehensive, but he did succeed in publishing an account of his trip in the local newspaper. Takara, the schoolteacher, couched his account in the idiom of a boys' own adventure story, no doubt hoping to inspire a younger generation with tales of scientific derring-do. He made much of the fact that he was headed off to explore a 'desert island' (*mujintō*), although the islands were in fact well known to Yaeyama islanders. And in language that echoed the sublime Romantic aesthetic deployed by Shiga Shigetaka, Tamaoki Han'emon and William Alanson Bryan, he described his first sighting of the islands as follows:

> dawn broke and at 7am we could see Minami-kojima hazily just below the horizon. 'So that is one of the uninhabited Senkaku Isles' I thought; 'it's just like I dreamed it would be!' The boat went up and down, shaking terribly, and huge waves broke against the sides of the boat and washed across the deck ... Through my binoculars I could see flocks of birds on Minami and Kita-kojima. Just looking at the cliffs and precipices of Uotsuri-shima's southern coastline made me shudder.[54]

Takara's sensationalised account was serialised in ten instalments over the course of a month.[55] It was so successful that it was picked up later the same year by the *Urusu Shimpō* on Okinawa Main Island – bringing this provincial schoolteacher on the periphery of the archipelago to the attention of an Okinawa-wide reading public for the first time.[56] Within a year, Takara was head of USCAR's agricultural improvement research institute. He was also appointed assistant professor at the newly founded University of the Ryukyus, despite possessing no doctoral training at the time. He would remain affiliated with the

[53] Takara Tetsuo, 'Kanshū no aisatsu' in Senkaku Shotō Bunken Shiryō Kenkyū Hensankai, *Senkaku kenkyū* 1, 1.

[54] Takara Tetsuo, 'Mujintō tanbō ki: 1', *Nanryū Taimusu* (25 April 1950) in Senkaku Shotō Bunken Shiryō Kenkyū Hensankai, *Senkaku kenkyū* 1, 51.

[55] Senkaku Shotō Bunken Shiryō Kenkyū Hensankai, *Senkaku kenkyū* 1, 14.

[56] OPA J 12 2 001–12, Bunkazai hogo iinkai iin shokuin rirekisho-shū: 'Takara Tetsuo rirekisho'.

institution for the rest of his life, eventually rising to the rank of university president.[57]

In 1952 Takara returned to the Senkaku Islands at the head of a much larger-scale expedition mounted by the Agriculture and Fisheries Ministries, with the aim of developing the economic resources of the islands. Takara was hopeful that the Senkaku group might prove to harbour commercially valuable phosphate deposits, and in his account of his initial trip he had broached the possibility of rebuilding the island's decimated seabird colonies so as to develop a sustainable guano mining industry.[58] The expedition departed with much fanfare, but the results were disappointing, for the islands lacked the coral and limestone base that produced the kind of high-grade phosphate to be found on Rasa Island and the neighbouring Yaeyama island of Haterumajima.[59] Neither did the surveys of the islands' fisheries and forestry resources succeed in spurring further investment in the islands.[60] Even the fish-processing plant that Takara had visited two years earlier had gone out of business, never to reopen.[61]

After the underwhelming results of this second mission, Takara's interests began to drift away from the Senkaku Islands for a while. He deepened his research into the habu, eventually earning the nickname 'Dr. Snake' for his efforts. And he led expeditions in search of rare fauna such as the Iriomote wildcat.[62] He also became an active member of the RCPPC, with special responsibility for the classification of fauna. His first involvement with the committee came when he was tasked with assessing the ecological implications of the reconstruction of Shuri Castle. By 1960 he had been promoted to specialist deliberator on the committee, and by 1965 he was responsible for investigating all five of the natural monuments designated in that year, many of which swiftly found their way onto commemorative Ryukyu

57 'Takara Tetsuo-san shiko – Ryūkyū Daigaku gakuchō', *Okinawa Taimusu* (19 May 2014).

58 Takara Tetsuo, 'Mujintō tanbōki' in Senkaku Shotō Bunken Shiryō Kenkyū Hensankai, *Senkaku kenkyū* I, 60.

59 Tanahara Seiichi, 'Senkaku no dosei, kaichō chōsa ni sanka shite', in Senkaku Shotō Bunken Shiryō Kenkyū Hensankai, *Senkaku kenkyū* 1, 305–6.

60 Kami Unten Kensei, 'Senkaku Rettō chōsa dōkō jisshū ki' in Senkaku Shotō Bunken Shiryō Kenkyū Hensankai, *Senkaku kenkyū* 1, 311.

61 Iramina, 'Chichi no ikotsu shūshū wa, jibun no fune to te de', 352.

62 For a history of Takara Tetsuo and the Iriomote wildcat, see Togawa Yukio. *Iriomote yamaneko* (Tokyo: Jiyū Kokuminsha, 1972).

government-issued postage stamps: the Yonaguni indigenous horse, the Kerama dwarf dear, the yellow-marginated tortoise and two types of native butterfly.[63] Around the same time Takara managed to capture a dugong, a rare manatee-like creature indigenous to the Southwest Indo-Pacific, on the island of Miyakojima. This species, too, was soon listed as a natural monument, although unfortunately the dugong itself died shortly after arriving in Naha.[64] Takara stuffed it, and his family feasted on its meat for weeks.[65]

The Origins of the Senkaku Nature Conservation Movement

Then, in 1963, came the letter from the Tokyo Meteorological Observatory asking about albatrosses on the Senkaku Islands. The enquiry therefore aroused an immediate flurry of interest, with the *Okinawa Times* announcing breathlessly that the Yamashina Institute, Japan's Agency for Cultural Affairs, or perhaps even the Maritime Self-Defense Force was planning an expedition to the islands.[66] When an expedition from the mainland failed to materialise, the bureaucrats and academics who staffed the Ryukyu Cultural Properties Protection Committee (RCPPC) decided to take matters into their own hands, despatching a ship across the choppy waters of the East China Sea in search of the newly minted natural treasure of the Japanese nation.[67]

Takara's intensive involvement in RCPPC, along with his prior knowledge of the Senkaku Islands, made him the obvious candidate to lead the expedition. He had long known that albatrosses were once plentiful in the islands, although he was unable to find any nesting sites on his early trips. The nearest he had come was during a 1953 expedition he had led with the university, during which one student had glimpsed a large white bird. Then again, tracking the species down had not been a primary objective of his earlier surveys. The 1963 expedition was to be an altogether more purposeful affair, setting out to scour all the islands for signs of albatrosses or their nesting sites. The Ryukyu Meteorological Agency also dispatched an oceanographer to

63 Ryūkyū Seifu Bunkazai Hogo Iinkai, *Bunkazai yōran 1965-nenban*, 26–7.
64 Senkaku Shotō Bunken Shiryō Kenkyū Hensankai, *Senkaku kenkyū* 1, 374.
65 Takara Tetsumi, personal conversation with author, June 2017.
66 'Senkaku Rettō no ahōdori', *Okinawa Taimusu* (29 January 1963).
67 Senkaku Shotō Bunken Shiryō Kenkyū Hensankai, *Senkaku kenkyū* 1, 7–8.

survey the islands' biogeography, and such was the public interest in albatrosses at the time that four Naha-based journalists accompanied the expedition, including a reporter from the Ryukyuan Broadcasting Corporation who hoped to telecast live from the islands.[68]

The expedition was for the most part a disaster. The boat was showered in warning tracer fire after straying into a stretch of water used by the US Air Force for bombing practice. Later, it capsized in rough water. No one was drowned, luckily, but the engine was soaked through, and by the time the expedition made landfall on the islands there was hardly any time left to conduct surveys. The attempted telecast failed. And most depressingly of all, not a glimpse was to be had of any albatrosses.

The only discovery of note was an unwelcome one. In the decade since Takara's last survey, the Senkaku Islands had come to be frequented by Taiwanese fishermen, who were in the habit of using simple rafts to traverse the reefs ringing the islands, and landing ashore to hunt for seabirds and their eggs (Figure 7.3).[69] The news made a splash not just in the Okinawan but also the mainland Japanese media, which denounced the 'inexcusable overhunting of the seabird flocks that cover out islands'.[70] Takara admitted that this was not the first time he had encountered Taiwanese fishermen collecting eggs on the islands, but he insisted that the scale of activity had increased dramatically since his last visit. Many of the fishing vessels, he claimed, were now coming for the express purpose of harvesting eggs. This 'indiscriminate slaughter' of seabirds, he argued, should not be permitted for a number of reasons. The islands' birds had not been sufficiently studied, but offered valuable clues as to the state of the neighbouring fishing grounds. Takara recommended that all known seabirds on the islands, and also any albatrosses that might happen to be discovered on them in the future, be placed under official protection, and that notices be erected on the islands forbidding unauthorised landing except in emergencies. Taiwanese fishermen in particular ought to be dissuaded in landing, through the offices of USCAR.[71]

The parallels with the Hawaiian bird panic of the 1900s are remarkable. A naturalist surveying islands abandoned by his own country's citizens had sounded the alarm that ingress by an alien other was

68 Senkaku Shotō Bunken Shiryō Kenkyū Hensankai, *Senkaku kenkyū* 1, 7–8.
69 Bukh, *These Islands Are Ours,* 153; Nishimura, 'Taiwan higashi kaigan ni okeru gyorō gijutsu no bunka shigenka', 40.
70 'Shimajima ōu kaichō no mure' *Asahi Shimbun* (21 May 1963).
71 Takara Tetsuo, 'Ahōdori o saguru', *Minami to Kita* 26 (March 1964): 168.

Figure 7.3 Taiwanese fishermen collecting eggs on the Senkaku Islands (c.1963–1968)
Source: Asahi Shimbunsha, *Okinawa no kotō* (1969)

threatening valuable wildlife, in the process attracting widespread media coverage. But the distinctive borderland politics of the East China Sea resulted in a different outcome. Sixty years earlier, William Alanson Bryan had been able to enlist support for Hawaiian bird conservation at the highest levels of the United States government. Campaigns to protect birds on the Senkaku Islands, however, secured no such support, and the RCPPC refrained from acting on Takara's recommendation.

Two years later, however, the RCPPC sent Takara to Nakanokamijima, an islet just off the Yaeyama island of Iriomote. Here, too, he discovered Taiwanese fishermen systematically culling the island's bird colonies. Nakanokamijima had already been designated a no-hunting zone but this was not enough, Takara insisted. Echoing the logic of Yamashina Yoshimaro and the WBSJ, he successfully argued that only designating the entire island as a natural monument would provide the necessary impetus to ensure its birds were protected.[72]

The year after the RCPPC granted Nakanokamijima natural monument status, Takara launched his fifth and final trip to the Senkaku Islands in July 1968. This time he reported evidence of even larger-scale hunting. Of the twenty-one Taiwanese fishing boats moored around the islands, at least five had their decks loaded up with eggs. Conversations with the fishermen revealed that the operations were distinctly commercial: the eggs were to be pickled, and the fledglings went to street venders to be grilled on sticks. Moreover, a band of more than a hundred Taiwanese workers had set up camp on the island – primarily for the purpose of salvaging scrap metal from a recent shipwreck, but many labourers were also helping themselves to seabirds.[73]

Once again Takara sounded the alarm, calling for USCAR to designate the islands as a bird reserve to be monitored by a regular patrol ship. And he did succeed in convincing the Ishigaki Police Department to despatch two constables to investigate his reports.[74] The Taiwanese fishermen on the islands professed bafflement, claiming that they had been visiting the islands to collect eggs since before the war, after all. The constable responded that before the war Taiwan had been part of Japan, but now it was a foreign country. The islands were therefore off-limits to them now.[75] (See Figure 7.4.)

By this time the political stakes had heightened substantially. The escalation of the Vietnam War had stoked Okinawan demands for reversion, and the looming renegotiation of the US-Japan Security Pact had become an obvious focal point for protestors. In December 1970, a drunken Okinawan man was hit by an American car near the Kadena Air Force Base. The incident sparked a spontaneous protest that quickly descended into violence, with five thousand Okinawans facing off against seven hundred military police officers: molotov cocktails

72 Takara Tetsuo, 'Ryūkyū Nakanokamijima no kaichō', *Yamashina Chōrui Kenkyū Hōkoku* 6 no. 1–2 (1970): 194.

73 Takara Tetsuo, 'Senkaku Rettō no kaichō ni tsuite' in Senkaku Shotō Bunken Shiryō Kenkyū Hensankai, *Senkaku kenkyū* 1, 233.

74 'Senkaku Rettō no kaichō ni tsuite', in Senkaku Shotō Bunken Shiryō Kenkyū Hensankai, *Senkaku kenkyū* 1, 9.

75 Niiro Tetsutarō, 'Beikaigun kansen wa sakusenka shika ugokasenai', in 'Senkaku Rettō no kaichō ni tsuite', in Senkaku Shotō Bunken Shiryō Kenkyū Hensankai, *Senkaku kenkyū* 1, 370.

Figure 7.4 Ishigaki police officers erect a sign forbidding unauthorised landings on the Senkaku Islands (1968)

were lobbed, teargas was fired, car were burned, twenty-one protestors were arrested, and over fifty American servicemen were injured.[76]

As reversion loomed closer, so did the dispute over the Senkaku Islands become more heated. In May 1968, the United Nations Economic Commission for Asia and the Far East (ECAFE) had announced, at a meeting in Tokyo, its intention of surveying the East China Sea's seabed for oil and gas, reviving interest in the area's economic potential.[77] And Chinese nationalists in Taiwan and Hong Kong framed the impending American handover of the islands to Japan as yet another instance of US imperialism. As hunger-striking

[76] Sarantakes, Keystone, 176; Masamichi S. Inoue, *Okinawa and the US Military: Identity Making in the Age of Globalization* (New York: Columbia University Press, 2017), 53–4.

[77] Reinhard Drifte, 'The Japan-China Confrontation Over the Senkaku/Diaoyu Islands – Between "Shelving" and "Dispute Escalation"', *Asia-Pacific Journal: Japan Focus* 12 issue 30 no. 3 (28 July 2014): 2; Barthélémy Courmont, 'Territorial Disputes and Taiwan's Regional Diplomacy: The Case of the Senkaku/Diaoyu/Diaoyutai Islands', *The Journal of Territorial and Maritime Studies* 1 no. 1 (2014): 120.

students in Hong Kong put it, 'The American and Japanese governments are now trying to seize part of our motherland'.[78] In September 1970, a group of Taiwanese newspaper reporters landed on the islands and hoisted the flag of the Republic of China. It was retrieved by the Ryukyu government and promptly emblazoned across the pages of newspapers in both Japan and Taiwan.[79] By the end of the year both Taipei and Beijing had asserted sovereignty over the islands, demanding that they be turned over to them to administer, not to Japan.[80]

Protestors in Okinawa and mainland Japan also began agitating over the issue. Koga Tatsushirō's son resurfaced, insisting that he had inherited the islands from his father and was still paying property tax on them.[81] Ryukyu government legislators filed a petition urging the US authorities to refute China's 'baseless claims' to the islands.[82] And islanders on Ishigaki suddenly remembered relatives who had drowned near the islands while being evacuated to Taiwan during the war, and insisted on erecting a stele to commemorate them.[83] The US found itself caught in the middle, wary of alienating either of its allies in Taipei or Tokyo and anxious not to let the issue derail the Nixon administration's ongoing attempt to seek détente with Beijing, either. When the Japanese government announced it would open a weather observatory on the islands after reversion, State Department officials scrambled to quash the plan lest it inflame matters still further.[84]

For the same reason USCAR's stance on Senkaku Island bird conservation swung from indifference to outright hostility. In March 1971, researchers from Nagasaki University announced confirmed sightings of Steller's albatrosses on the small Senkaku islands of Minami and

78 OPA Office of East Asia and Pacific Affairs, Asian Communist Affairs (ACA), Subject Files, 1963–72 (DVD 47 pdf 2012/3): 'Senkaku protestors send letters to Chinese leaders'.

79 Unryu Suganuma, *Sovereign Rights and Territorial Space in Sino-Japanese Relations: Irredentism and the Diaoyu/Senkaku Islands* (Honolulu: University of Hawai'i Press, 2000): 132; Bukh, *These Islands Are Ours*, 132.

80 Hara, 'Okinawa, Taiwan, and the Senkaku/Diaoyu Islands', 10.

81 OPA, Office of East Asia and Pacific Affairs, Japanese newspaper clippings re: Kuomintang: 'Senkaku Shotō wa watashi no mono' (6 September 1970).

82 OPA R 09 1 004–18 'Dai 42 teirei ketsugi antei – gun kankei tokubetsu iinkai'.

83 Iramina, 'Chichi no ikotsu shūshū wa, jibun no fune to te de', 360–1; Senkaku Shotō Bunken Shiryō Kenkyū Hensankai, *Senkaku kenkyū: Takara gakujutsu chōsadan shiryōshū* 2 (Naha: Dētamu Rekiosu, 2007b): 267.

84 'US killed '71 plan for Senkaku Islands weather station', *Japan Times* (5 September 2013).

Figure 7.5 Ryukyu Post Office stamp portraying Steller's albatrosses (issued 14 April 1972)

Kita-kojima.[85] By this time American officials were so nervous about anything Senkaku-related that to designate the species a Ryukyu natural monument was out of the question. But Okinawan conservationists did one mount one final effort to inscribe the Senkaku Islands and their albatrosses onto the (soon to be Japanese) body politic. In April 1972, only a month before reversion, the Ryukyu Post Office issued the penultimate postage stamp before it was shut down (Figure 7.5). The stamp was issued under the title 'Seabirds, Sea and Islands', but knowledgable twitchers would have spotted that the birds depicted on the stamp were clearly albatrosses. Indeed, the artist who

[85] 'Ikite ita "maboroshi no tori": Chōdai chōsa tai ga kakunin', *Nagasaki Shimbun* (21 March 1971).

designed the stamp used a stuffed albatross displayed in the lobby of the Naha post office building as his life model. Later he would recall that 'only by camouflaging the theme of the stamp were we able to pass it successfully through the stringent checks carried out by USCAR'.[86]

As a gesture the stamp was both sly and (quite literally) tokenistic, clearly meant for domestic consumption within Japan – and by a rather select community of bird-fancying philatelists at that. But it was also a defiant assertion of the irrevocable three-way link between a nation, its biodiversity and its territory. Thongchai Winichakul has described the way in which various 'natural features' such as mountains, rivers and watersheds are appropriated in order to construct the 'geobody' of the nation – and here, as with Torishima, we see Steller's albatross mobilised as a charismatic, rare and uniquely Japanese species to argue that the Senkaku Islands are an integral part of the Japanese nation state.[87]

Takara Tetsuo's Legacy Lives On

In 1978, as part of the normalisation of Sino-Japanese relations, the Senkaku sovereignty dispute was shelved. Deng Xiaoping famously quipped that a resolution to the dispute was best left for later, wiser generations to resolve.[88] But in recent years the Senkaku nature conservation movement has seen something of a revival. There are several reasons for this. One driver has been the territorialisation of maritime space under international law, which has increased the potential economic importance of remote islands. In 1994, after more than two decades of negotiation, the United Nations Convention on the Law of the Sea (UNCLOS) finally entered into force, and Japan promptly declared a 200 nautical mile exclusive economic zone around its coastline, including that of the Senkaku Islands.[89] Just as importantly, nationalist groups in both Japan and China have sought to publicise

86 Senkaku Shotō Bunken Shiryō Kenkyū Hensankai, *Senkaku kenkyū* 2, 246.

87 Thongchai Winichakul, *Siam Mapped: A History of the Geo-Body of a Nation* (Honolulu: University of Hawai'i Press, 1997), 132.

88 Zhongqi Pan, 'Sino-Japanese Dispute Over the Diaoyu/Senkaku Islands: The Pending Controversy from the Chinese Perspective', *Journal of Chinese Political Science* 12 no. 1 (2007): 74.

89 Erica Strecker Downs and Phillip C. Saunders, 'Legitimacy and the Limits of Nationalism: China and the Diaoyu Islands', *International Security* 23 no. 3 (Winter, 1998–1999): 124. As Bukh notes, expert opinion is split as to exactly

the issue as a way of elevating patriotic consciousness among their respective citizenries.

The catalyst for the most recent flurry of attention was a collision between a Chinese fishing trawler and a Japanese coastguard vessel in September 2010. The coastguard detained the vessel's captain, prompting Beijing to suspend diplomatic relations until they were released. The Japanese government endured a hail of criticism for its handling of the crisis, and Prime Minister Kan's decision to release the Chinese captain without charge was one factor in his party's defeat in the December 2012 elections.[90]

As right-wing Japanese nationalists rallied to make the cause for defending their country's claim to the Senkaku Islands, nature conservation emerged as a key theme. This time the immediate focus of attention was not Steller's albatross but an even more obscure species: the Senkaku mole (*Mogera uchidai*). Remarkably little is known about the Senkaku mole. The sum accumulation of evidence for the creature's existence consists of one specimen collected in 1979, as well as a 1991 photograph of a hole in the ground that may or may not be a mole's burrow.[91] The first calls to conserve the species came in the late 1990s, after an all-Japan conference of mole scientists issued an 'urgent appeal to conserve the natural environment' from marauding goats that were eroding Uotsuri-shima's topsoil, threatening the mole and other obscure indigenous biota.[92] But it was not until the fishing boat collision that the campaign began to gain any substantive political traction.

In October 2010, only a month after the arrest of the Chinese fishing captain, a celebrity mountaineer-turned-environmentalist by the name of Noguchi Ken founded the Society to Protect the Senkaku Mole. The society circulated a petition calling on the national government to permit an ecological survey expedition to the islands to ascertain the

how valuable the East China Sea mineral deposits actually are. See Bukh, *These Islands Are Ours*, 130.

90 Paul Midford, 'The Influence of Public Opinion on Foreign Policy in Asia: The Case of Japan', in *The Sage Handbook of Foreign Policy Vol. II*, ed. Takashi Inoguchi (Los Angeles: SAGE Reference, 2019): 388.

91 IUCN Red List 'Senkaku Mole' (14 December 2015): www.iucnredlist.org/species/14563/22323327. Last accessed 6 July 2020.

92 Yokota Masatsugu and Yokohata Yasushi, 'Senkaku Shotō Uotsuri-shima no yaseika yagi mondai ni tsuite', *Wildlife Conservation Japan* 5 no. 1–2 (2000): 10.

(apparently dire) status of the benighted mole, and formed a sister organisation of Diet members headed by future Tokyo Mayor Koike Yuriko. It even managed to design its own mascot, 'Mogumogu-sen chan'.

A sceptic might wonder how many members of the Senkaku Mole Protection Society had even heard of the species even a month earlier.[93] But a number of Japanese scientists did decide to throw in their lot with the movement, lending it a veneer of credibility. One of these was Niiro Yoshima, a former student of Takara Tetsuo. Niiro had participated in several of Takara's Senkaku surveys, as well as the Nagasaki University expedition that had eventually spotted Steller's albatrosses nesting on the islands. And his later career followed very much in Takara's tracks: he went on to teach at the University of Ryukyus and to serve on the RCPPC's post-reversion incarnation, the Okinawa Prefectural Cultural Property Protection Committee.[94]

In January 2011 a now retired Niiro gave a speech in Ishigaki on the need to 'protect the unique flora and fauna of the Senkaku Islands'. As a botanist, Niiro's primary concern was goat eradication, but he also mentioned Steller's albatross, which he pointed out was officially recognised as a rare wild species by the national government, and which 'in the whole world breeds only in the Izu and Senkaku Islands'. He called on the national government to develop a policy aimed at protecting the birds' breeding grounds – a plea that harked back to Takara's campaigns in the years leading up to reversion. (In 2017 I wrote to Professor Niiro, who was by then more than ninety years old. He declined to be interviewed, but did issue an unequivocal statement condemning Chinese attempts to 'rob'[95] Japan's sovereign territory.)

Another key ally was Yokohata Yasushi, a professor of ecology at Toyama University. Yokohata has been campaigning to save the Senkaku mole since 1997, for reasons not obviously connected to nationalist politics. Earlier in his career he called for international collaboration between researchers and NGOs from the relevant

93 Yasushi Yokohata, Yukio Ikeda, Masatsugu Yokota and Hideharu Ishizaki, 'Effects of Introduced Goats on the Ecosystem of Uotsuri-Jima in the Senkaku Islands, Japan, as Assessed by Remote-sensing Techniques', *Biosphere Conservation* 5 no. 1 (2003): 39–46.
94 Senkaku Shotō Bunken Shiryō Kenkyū Hensankai, *Senkaku kenkyū 1*, 367.
95 *ubaitoru.*

countries (i.e. Japan, China and Taiwan) in order to eradicate the goats on the island.[96] But these calls had largely fallen on deaf ears, and by 2010 he was ready to cast in his lot with Noguchi and his allies in the Liberal Democratic Party. In June 2020 Professor Yokohata was kind enough to speak to me about his involvement with the Senkaku Mole Protection Society. The society's motivations, he assessed, were 'half scientific, half political'. But if it had not been for him, he admitted, it might have been 'almost wholly political'. He recalled being surprised to discover that Noguchi, the celebrity mountaineer, had once been a member of the Japan Youth League, a nationalist organisation that had been campaigning on the Senkaku issue since the 1970s. (Indeed, it was the Japan Youth League that had first introduced goats to Uotsuri-shima in 1978 as part of an ill-conceived attempt to provide food supplies for a lighthouse they were building on the island.)[97]

The Senkaku nature conservation campaign escalated dramatically in April 2012 when Ishihara Shintarō, the mayor of Tokyo, announced that he would solicit public donations so that the metropolitan government might purchase the islands from their putative owner, who had bought the rights to them from Koga's son in the 1970s.[98] If the original annexation of Japan's desert islands was driven by commercial entrepreneurs such as Tamaoki Han'emon, Mizutani Shinroku and Koga Tatsushirō, Ishihara was a classic example of what Alexander Bukh calls 'national identity entrepreneurs': 'signifying agents who engage in linking a certain issue or event with the nation or the national interest'.[99] Ishihara, a well-known supporter of various right-wing nationalist causes, had a prehistory of Senkaku-related activism. Indeed, in his youth he had been a founding member of the Japan Youth League. And he cultivated a reputation as a maverick and upstart challenging the complacent centrism of post-war Japanese elites. He took great relish in blindsiding the political establishment with his plan to buy the islands, which he announced in an almost

[96] Yokota and Yokohata, 'Senkaku Shotō Uotsuri-shima no yaseika yagi mondai ni tsuite', 9.
[97] Yokohata Yasushi, interview with author, June 2020.
[98] Antoni Slodkowski, 'How debts and double-dealing sparked Japan-China islets row', *Reuters* (12 November 2012): www.reuters.com/article/us-china-japan-family-idUSBRE8AA0EY20121111. Last accessed 1 April 2022.
[99] Bukh, *These Islands Are Ours*, 10.

offhand way towards the end of a rambling speech at the right-wing Heritage Institute think-tank in Washington, DC.[100]

During this first speech, Ishihara did not seem to have a particularly clear idea of what he would actually do after he bought the islands – apart from 'protect' (*mamoru*) them. But it was obvious that Ishihara's real motivation was to use them as a stage on which to perform nationalist agitprop, whipping up public passions over the territorial dispute to the point where the central government felt compelled to take a more assertive stance against China. The question was, what form would this agitprop take?

Nature conservationists quickly flocked to Ishihara's cause. The Senkaku Mole Protection Society, for instance, urged its members to donate to his fund-raising drive to help buy the islands. Another early supporter was the mayor of Ishigaki, Nakayama Yoshitaka, who met with Ishihara and Tokyo government officials only a week after the Heritage Foundation speech. The two local governments then issued a press release committing to 'cooperate on nature conservation and ecological surveying' on the islands.[101] As Ishihara's proposal for managing the islands developed over the ensuing months, the environmental management aspect became increasingly prominent. The Tokyo government published a report that compared the Senkaku Islands with the Bonin Islands, since 2011 a UNESCO-accredited world natural heritage site. The Bonins had once been uninhabited, the report observed – just as the Senkaku were now. But enlightened governance had succeeded in developing the islands sustainably, for instance by eradicating invasive goats. Tokyo, therefore, possessed the wherewithal to transform the Senkaku into 'a second Bonin Islands . . . by protecting the environment whilst also securing the livelihoods (*seikatsu*) of the people in the area'.[102]

In 2012 Ishihara then applied for central government permission to land a survey team on the islands. The request was denied, but in September he nevertheless despatched a team to conduct an offshore

[100] Ishihara Shintarō, speech delivered at The Heritage Foundation, Washington, DC (16 April 2012): www.youtube.com/watch?v=9DenbvnNlvI&t=2785s. Last accessed 6 July 2020.

[101] 'Ishigaki-shichō, Senkaku kōnyū ni 'sandō' Ishihara-tochiji to kaidan', *Nihon Keizai Shimbun* (23 April 2012).

[102] Yamada Yoshihiko, *Senkaku Shotō no kanri mondai ni kan suru hōkoku* (Tokyo: Tōkyō-to, 31 March 2013), 3–4.

survey. It became a media circus. With planes and helicopters chartered by newspapers and TV stations buzzing overhead, the team circled the islands in a dirigible, at times skirting the shoreline from a distance of only a few meters. Video footage from the survey played by Sakura TV, a right-wing satellite and online news channel, lingered on the islands' colonies of seabirds but also on the litter that had accumulated on the shoreline and the goats that roamed the steep hillsides, allegedly stripping away foliage and causing soil erosion. Ishihara was so pleased with the results that he announced that he would personally lead a second expedition to the islands the very next month, and that this time he would flout the government's ban on making landfall.[103]

The prospect of Ishihara securing the Senkaku Islands as a permanent stage for his theatrics so alarmed Japan's centre-left ruling coalition that, only three days after the expedition, it announced that it would purchase the islands itself.[104] If the government's plan was to defuse international tensions then it backfired, for Japan's 'nationalisation' (*kokueika*) of the islands provoked a furious response from Beijing. In that sense, if Ishihara's strategy had been to focus public attention on the Senkaku issue then it had worked perfectly. Even after his bid to purchase the islands was foiled, he penned a tract urging the central government to administer the islands more proactively so as shore up Japanese sovereignty claims. 'If the islands had become the property of the Tokyo metropolitan government, we could have built facilities that would allow fishermen to operate safely – harbours, lighthouses, wireless relay stations, and so on. This would make it possible to administer the surrounding waters more effectively.' The central government, by contrast, 'has not even surveyed the islands since 1975, for fear of offending China'.[105]

Nakayama also attempted to keep up the drumbeat of pressure on the central government. In December he published a book arguing that

103 'Senkaku Shotō o to ga kaijō chōsa' (3 September 2012): www.youtube.com/watch?v=GWKOqUR7mkg. Last accessed 6 July 2020; Katsuragi Nami, 'Senkaku repōto – Tōkyō-to Senkaku Shotō yōjō chōsa', *SakuraSoTV* (6 September 2012).

104 'Senkaku, kuni ga kōnyū de gōi 20.5 oku en tochiji ni mo dentatsu.' (5 September 2012): www.asahi.com/special/minshu/TKY201209040777.html. Last accessed 6 July 2020.

105 Nakayama Yoshitaka, *Chūgoku ga mimi o fusagu Senkaku Shotō no futsugōna shinjitsu: Ishigaki shichō ga tsutsuguru Nihon gaikō no aru beki sugata* (Tokyo: Wani Bukkusu PLUS Shinsho, 2012), 6.

Japan must defend its claims to the Senkaku group for a number of reasons – economic and strategic, yes, but also out of a sense of responsible environmental stewardship. Japan had a duty to protect unique species such as the Senkaku mole and river crab, and also to prevent the islands from becoming littered with garbage 'that is thought to have drifted from foreign countries'.[106] In a February 2013 interview with the conservative-leaning *Sankei Shimbun* Nakayama went one further, calling on the central government to revise its (then-pending) application for UNESCO natural heritage status for Okinawa, so as to include the Senkaku Islands as well.[107] In attempting to solicit support from international institutions dedicated to nature conservation, Nakayama was in many ways building on strategies honed by earlier generations of conservationists in Japan and Okinawa.

This, then, was the context within which Central China Television warned, in the autumn of 2013, of a Japanese plot to reclaim the Senkaku Islands using the 'pretext' of nature conservation. The grist for this claim was an announcement by the deputy director of the Yamashina Institute the previous day, related to an effort to establish a new albatross colony on the island of Mukojima in the Bonin group, as an insurance policy in case the Torishima breeding population were wiped out by another eruption. This project had been ongoing since 2008, with mixed success: the majority of eggs laid in the artificial colony have been infertile. Recent genetic analysis, the institute announced, suggested that this was due to the fact that the Torishima-descended birds on Mukojima had been mating with birds descended from the Senkaku albatross population, which in genetic terms are distinct enough to result in outbreeding.[108] This discovery potentially had large ramifications for the future success of the Mukojima nesting site, for it suggested that the two albatross populations ought to be considered as, if not separate species, then certainly distinct 'conservation units' that would need to be somehow kept separate in order to maximise the yield of fertile eggs.

All of which is to say that the press release primarily focused on the nesting site on Mukojima, and only secondarily on the Senkaku Islands.

[106] Nakayama, *Chūgoku ga mimi o fusagu Senkaku Shotō no futsugōna shinjitsu*, 60.

[107] '"Senkaku" mo sekai isan tōroku o: Ishigaki-shi, kuni ni genchi chōsa yōkyū e', *Sankei Shimbun* (4 February 2013).

[108] 'Haha ahōdori wa Senkaku umare, Ogasawara Mukojima'.

In fact, the wording of the institute's press release was rather careful, simply noting that on the Senkaku Islands 'the current situation makes it difficult to survey the birds' breeding status or conduct conservation activities'. But when interviewed by Japanese media outlets the Yamashina Institute's deputy director, Ozaki Kiyomasa, went one further. 'Quite apart from political factors,' he told the *Nihon Keizai Shimbun*, 'it is necessary to protect the birds [on the Senkaku Islands]. We're hoping that we'll be able to conduct an on-the-ground survey.'[109]

Now, for CCTV to accuse Ozaki of a cynical plot to undermine Chinese sovereignty through the masquerade of science is, in one sense, ludicrous. There are moments when a researcher must reveal his positionality, and here I should say that the staff of the Yamashina Institute showed considerable kindness in assisting me as I researched this book. Like Professor Yokohata, they struck me as sincere and dedicated individuals to a fault, defined if anything by their somewhat otherworldly passion for birds. I am inclined to believe that their motives are for the most part quite distinct from those of Ishihara, Nakayama and the mole-crusading faction of the Liberal Democratic Party.

On the other hand, Ozaki's contention that nature conservation on the Senkaku Islands could be separated from politics was, to put it mildly, optimistic. To call for the Japanese government to conduct on-the-ground survey of the Senkaku islands only a year after Ishihara's media circus was almost bound to be inflammatory. And as this study has shown, across the twentieth history campaigns to conserve the nature of bird islands have often gained political traction when they have overlapped with concerns to conserve sovereignty over the islands themselves. Is it possible to imagine another way to conserve the moles and albatrosses of the Senkaku Isles, to envisage a form of environmental diplomacy in which Chinese and Japanese sharpshooters work together to pick off the goats that roam the cliffs of Uotsuri-shima?

I put this question to Professor Yokohata, who in his own early writings had gestured to such a possibility. He shook his head sadly: 'Under the present circumstances it does not seem likely.' In fact, as Sino-Japanese relations have improved from the nadir of 2012, the issue of nature conservation on the disputed islands has once again dropped off the political agenda. The Senkaku Mole Protection Society

[109] 'Haha ahōdori wa Senkaku umare, Ogasawara Mukojima'.

has not held a meeting since 2014.[110] Ishihara Shintarō died in 2022, and Noguchi the mountaineer has moved on to other challenges. Koike Yuriko is now Mayor of Tokyo, and is also apparently too busy to devote attention to the issue. With his hopes of making landfall on the islands dashed for now, Professor Yokohata is currently designing a submarine that can scour seawater for trace amounts of mole urine.

110 Senkaku Mogura o Mamoru Kai, 'Senkaku Mogura o Mamoru Kai no o-shirase (n.d.): www.senkaku-mogura.jp. Last accessed 6 July 2020.

Epilogue
Islands and Oceans

In October 2018, a Japanese travel writer made a startling discovery along the Hokkaidō coastline: an entire island had disappeared. In four years since the last cartographic survey, Esanbe Hanakita Kojima had vanished completely, presumably eroded either by wind or seasonal drift ice floating down from the Sea of Okhotsk. Even in its prime the island had been barely larger than a baseball stadium, but its disappearance attracted global news coverage. And if international news outlets treated the discovery as a humorous oddity, the tone of the Japanese coverage was more ominous. As the *Asahi Shimbun* pointed, out, 'if the island indeed no longer exists then our national maritime territory has shrunk'.[1]

The anxiety about the disappearance of Esanbe Hanakita Kojima was less about the island itself than about its relationship with the ocean surrounding it. Of course, islands have always been inextricably tied to the oceans. In sheer geological terms, many bird islands are composed of coral beds or fossilised marine microorganisms. Some have – much like Esanbe Hanakita Kojima – emerged from or been swallowed by the sea within the span of historical time. And ecologically, bird island ecosystems consist of tightly interlinked food webs that extend above and below the waves: the birds that nest onshore feed on fish and squid caught within a wide ocean radius, and themselves generate trophic cascades of faecal runoff that sustain plankton and the host of sea creatures that feast upon them.[2]

1 'Umi ni kiete shimatta? Hokkaidō no kojima, ryōkai ga semaru osore', *Asahi Shimbun* (31 October 2018); Justin McCurry, 'Japanese Island Disappears, without Anyone Noticing', *The Guardian* (2 November 2018); Brigit Katz, 'How a Japanese Island Quietly Disappeared', *Smithsonian Magazine* (2 November 2018); 'Hokkaidō no mujintō "Esanbe Hanakita-kojima" shōshitsu: Kaiho ga kakunin', *Nihon Keizai Shimbun* (24 September 2019).

2 A. N. Golovkin, 'The Effect of Colonial Seabirds on the Development of Phytoplankton', *Oceanology* 7 (1967): 521–52.

Human access to bird islands is also mediated by oceans. For millennia the Pacific Ocean kept Steller's albatrosses safe from land-borne predators – until it didn't. In the nineteenth century the 'low-friction terrain' of the ocean allowed bands of hunters to roam widely across vast swathes of pelagic space with comparative ease, and runways constructed during the 1930s and '40s have rendered some islands accessible by air as well. Still, even with the latest technology landings remain hostage to the whims of oceanic weather systems.

The geostrategic links between islands and oceans run similarly deep. In the imperial imagination dreams of colonising desert islands intermingled with broader visions of *Nan'yō* empire, naval supremacy and control of pelagic fisheries.[3] We have already seen how, in the 1930s, the Japanese Navy bankrolled fisheries operations to bolster its claims to the Spratly Islands. Only a few years ago, a security think tank accused China of adopting similar tactics, deploying what it ominously called a 'maritime militia' to defend atolls in the South China Sea.[4] Coastguards have also come to play an increasing role in island disputes. Ostensibly civilian organisations set up to rescue sailors and enforce customs duties, since the end of World War II they have become increasingly militarised. Some Japan Coast Guard vessels now come equipped with 20-millimetre gatling guns, the same weapons that are mounted on US fighter planes.[5] At one point Japanese and Taiwanese coastguard ships even opened fire on each other, albeit only with water cannons.[6]

Although islands have long been geologically, ecologically, economically and strategically linked to oceans, they are now *juridically* linked to them also. Maritime law was originally founded on Hugo Grotius' classic distinction between land as enclosed, sovereign space and the sea as free, non-sovereign space, and until 1994 states were only permitted to claim a narrow fringe of coastline as their territorial waters. But UNCLOS now grants states the right to claim exclusive

[3] Schencking, *Making Waves*; Tsutsui, 'The Pelagic Empire'.

[4] Conor M. Kennedy and Andrew S. Erickson, 'China's Third Sea Force, The People's Armed Forces Maritime Militia: Tethered to the PLA', *China Maritime Studies Institute China Maritime Reports* 1 (2017).

[5] Richard J. Samuels, '"New Fighting Power!" Japan's Growing Maritime Capabilities and East Asian Security', *International Security* 32 no. 3 (Winter 2007/2008): 100.

[6] 'Japan, Taiwan Duel with Water Cannon off Diaoyu Islands', *South China Morning Post* (25 September 2012). Last accessed 7 July 2020.

economic zones projecting up to 200 nautical miles from their coasts, and continental shelves projecting up to 350 nautical miles. This means that remote oceanic islands now potentially generate claims to vast tracts of sea and seabed containing valuable stocks of fish and minerals. A hundred years ago, prospectors claimed bird islands so as to extract their terrestrial resources. But now the islands' economic value lies offshore, even if the exact extent of these riches remains uncertain. As well as oil and gas, a Japanese government-funded oceanographic research group recently claimed (somewhat gnomically) that Japan's exclusive economic zone contains 'semi-infinite' supplies of the rare earths used to manufacture batteries and screen displays.[7]

This new legal linkage between islands and their surrounding oceans has given rise to new ways of thinking about the nation itself. The Japanese government is fond of boasting that when the maritime economic zones of these islands are factored in, the country ranks as sixth largest in territorial size in the world.[8] Japan's bird islands are so tiny that they would be invisible on any large-scale projection of the nation's territory, but one popular map produced by the Japan Coast Guard portrays them at magnified size so as to emphasise their significance as anchors of pelagic sovereignty (Figure 8.1).[9] Such representations have helped 'national identity entrepreneurs' such as Ishihara Shintarō and Noguchi Ken portray tiny, uninhabited islands as integral components of the Japanese geobody – as synecdoche for the nation itself.

This helps to explain the number of media column inches devoted to vanished Esanbe Hanakita Kojima. The island lay only 500 metres offshore, so its erosion has shaved but the merest sliver from Japan's exclusive economic zone. But news of its disappearance would have reminded many readers of another island that lies much further out to sea: the North

[7] Yutaro Takaya, Kazutaka Yasukawa *et al.*, 'The Tremendous Potential of Deep-Sea Mud as a Source of Rare-Earth Elements', *Nature Scientific Reports* (10 April 2018): www.nature.com/articles/s41598-018-23948-5.pdf. Last accessed 21 April 2022.

[8] Shinzō Abe, 'Speech by Prime Minister Shinzo Abe at the Grand Opening Ceremony for Special Events in Commemoration of the 20th Marine Day' (20 July 2015): https://japan.kantei.go.jp/97_abe/statement/201507/1212117_9924.html. Last accessed 21 April 2022.

[9] Kaijō Hoanchō Kaiyō Jōhōbu, 'Nihon no kaiyō tōgai nenzu' ([n.d.]): www1.kaiho.mlit.go.jp/JODC/ryokai/ryokai_setsuzoku.html. Last accessed 21 April 2022.

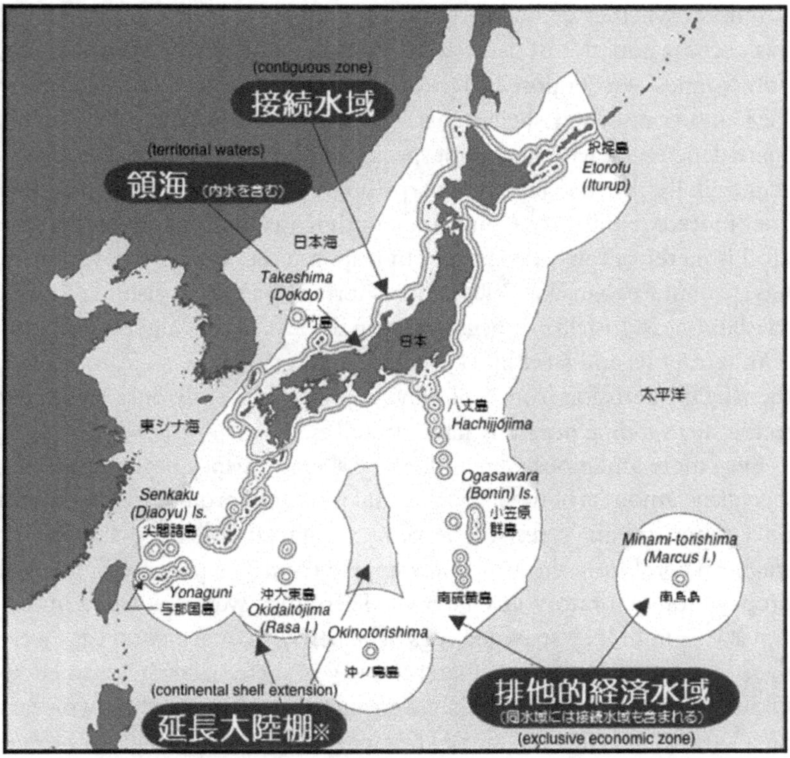

Figure 8.1 Map published by Japan Coast Guard to show Japan's oceanic territorial rights
Source: Japan Coast Guard website

Pacific atoll of Okinotorishima. Even compared to the other islands discussed so far, Okinotorishima has only the most fleeting history of human habitation. Annexed to Japan in 1931, in 1939 the Imperial Japanese Navy began building a lighthouse and weather observatory atop concrete blocks dropped onto the atoll's coral reef. But the project was abandoned upon the outbreak of the Pacific War, and was not resumed until as recently as 2007. At high tide even the highest elevations on Okinotorishima protrude mere centimetres above sea-level.[10]

[10] Yann-huei Song 'Okinotorishima: A "Rock" or an "Island"? Recent Maritime Boundary Controversy between Japan and Taiwan/China', in *Maritime Boundary Disputes, Settlement Processes, and the Law of the Sea*, ed. Jon M. van Dyke (Leiden: Martinus Nijhoff, 2009): 147–8.

Indeed, whether or not Okinotorishima even qualifies as an 'island' has become a matter of fierce controversy. For according to UNCLOS, only 'islands' – as opposed to 'rocks' or 'low-tide elevations' – generate EEZ and continental shelf rights. Control over vast tracts of oceanic space depends on whether or not specific atolls or cays can be regarded as islands in legal terms. Japan, unsurprisingly asserts that Okinotorishima does indeed qualify, but China and South Korea both maintain that the atoll is merely a low-tide feature. In response the Japanese government has spent more than US $200 million shoring up Okinotorishima's claim to island-ness. In 1987, it began ringing the atoll's two highest outcrops with tetrapods and steel blocks smothered in concrete, so as to protect them against erosion from typhoons. They are currently only connected to the surrounding ocean by narrow sluices.[11]

Even more ambitiously, in 2006 the Fisheries Agency began funnelling increasing amounts of money into a trial project to promote coral growth on Okinotorishima's reefs. The project entailed collecting samples of staghorn coral from the atoll and shipping them to a purpose-built coral propagation laboratory in Okinawa. There they were cultivated in seawater vats until they spawned eggs. The newly spawned coral eggs were then incubated for approximately one year before being returned to Okinotorishima. From there they were cultivated for a further two years on an artificial reef, before finally being deposited on existing reef surface encased in protective meshed coops. The results of this project were mixed. A 2016 analysis reported a coral survival rate of under 30 per cent, but the researchers claimed to have successfully transplanted around 100,000 coral colonies, and to have recorded instances of transplanted colonies increasing up to fourfold in area within two years.[12]

Still, even if this coral cultivation project does succeed in strengthening the reef around Okinotorishima, the legal implications of the intervention are unclear. UNCLOS stipulates that an island must be a 'naturally formed area of land'. So do Japan's various activities on and around Okinotorishima constitute interference in 'natural' processes of erosion and subsidence, or perhaps even an attempt to geo-engineer an 'island'

[11] Song, 'Okinotorishima: A "Rock" or an "Island"?, 149, 174–5; Clive Schofield, 'Sea Level Rise and Options to Secure Maritime Jurisdictional Claims', *Carbon & Climate Law Review* 3 no. 4 (2009): 412.

[12] Nakamura et al, 'Okinotorishima hatsu sango no hozen saisei II – gijutsu kaihatsu no shinpo ni tsuite', in *Midoriishi* (27) 5–9 (2016): 6–8; Correctiv.org, 'Japan is Growing an Island' (28 July 2017).

where before there was none? Or are they an attempt to conserve a fragile ecosystem that is threatened by anthropogenic climate change? Some Japanese lawyers have even begun to make common cause with the governments of low-lying island nations such as the Maldives, arguing that even if atolls are submerged by rising sea levels the territorial rights they generate should remain intact. Such arguments raise the prospect of ghost islands, that will survive as legal fictions even after they have been swallowed by the waves.[13]

Okinotorishima is not the only atoll whose island-ness is debatable according to UNCLOS. As we have seen in the previous chapters, Itu Aba (Ch: Taiping-dao) in the South China Sea has been the site of numerous failed guano mines, and was at one time a covert Japanese submarine refuelling station. Since World War II it has been administered by the Taiwanese government, which has essentially inherited the prewar Japanese claim to the Spratly group. Governments in both Taipei and Beijing use the 'island' of Itu Aba to make expansive claims over the waters of the South China Sea.

But in 2013 the government of the Philippines brought a suit to the Permanent Court of Arbitration at the Hague claiming that Itu Aba was, in legal terms, no more than a 'rock'. The atoll, the Philippines' lawyers argued, failed to meet another UNCLOS criterion of island-ness: that the land feature be capable of supporting 'human habitation and economic life'. This stipulation contains within it the kernel of earlier nineteenth-century formulations of international law: that sovereignty over territory is generated through continuous occupation of it. As we have seen, though, habitability is a slippery thing. Although the terms of UNCLOS emphasise physical geography, the tribunal admitted in its final ruling that the barrage of soil analyses and groundwater assessments provided by lawyers on both sides were not enough to conclusively determine whether or not Itu Aba was habitable.[14]

Instead, the court was forced to fall back on the historical record. And in an oddly recursive fashion, Taiwanese lawyers drew on earlier accounts by Japanese colonial boosters in order to argue that Itu Aba was a fit site for human habitation. The Rasa Company's guano mining

[13] Moritaka Hayashi, 'Sea-Level Rise and the Law of the Sea: Future Options', in *The World Ocean in Globalisation: Climate Change, Sustainable Fisheries, Biodiversity, Shipping, Regional Issues*, eds. Davor Vidas and Peter J. Schei (Leiden: Martinus Nijhoff, 2011): 187–206.

[14] UNCLOS Permanent Court of Arbitration Case No. 2013-19 (2016): 251.

operation on the atoll was cited as proof that it could support economic life, even though that mine had itself largely been a performative exercise. Accounts by Rasa Company-sponsored journalists that puffed up Itu Aba's agricultural potential were now recycled as proof that the atoll was indeed habitable.[15] Performance was layered atop performance, now in an effort to claim not the island but the waters that surrounded it. (The tribunal did not ultimately find these arguments convincing, and awarded the case in the Philippines' favour.)

Nature Conservation and the Territorial Temptation

Nature conservation has also served as one means of forging new legal connections between islands and oceans. In the past few years pelagic states have established a series of vast oceanic reserves anchored by bird islands. In 2006, nearly a century after Theodore Roosevelt established the Hawaiian Islands Reservation, President George W. Bush sat down to watch a documentary filmed by the oceanographer-adventurer Jean-Michel Cousteau on and around the islands of Midway and Laysan. Bush was so moved by the documentary that he decided to expand the bird reservation into the Papahānaumokuākea Marine National Monument (Figure 8.2), within which 'fishing, mining, oil exploration or other commercial activity' would be limited so as to protect a 'unique habitat to hundreds of rare species of birds and fish'.[16] A decade later President Barack Obama issued another executive order, expanding Papahānaumokuākea up to the limit of the islands' exclusive economic zone, banning all commercial fishing and making it, at 1,508,870 square kilometres, America's largest nature reserve.[17]

A number of things could be said about such gestures. For one thing, part of Papahānaumokuākea's purpose is to acknowledge Indigenous

[15] Ta-Wei Chang, 'Water Quality and Agricultural Environment Survey-Groundwater Quality and Hydrology Survey Report' (12 March 2016); Zueng-Sang Chen, 'Brief Report of Soil Resources Survey on Taiping Island' (14 Mar. 2016); Chinese (Taiwan) Society of International Law 'Amicus Curiae Submission' (23 March 2016) Exhibit 3.

[16] John M. Broder, 'Bush to Protect Vast New Pacific Tracts', *New York Times* (5 January 2009); Alison Rieser, 'The Papahānaumokuākea Precedent: Ecosystem-Scale Marine Protected Areas in the EEZ', *Asian-Pacific Law & Policy Journal* 13: 2 (2012).

[17] 'About Papahānaumokuākea': www.papahanaumokuakea.gov/new-about/. Last accessed 8 March 2018.

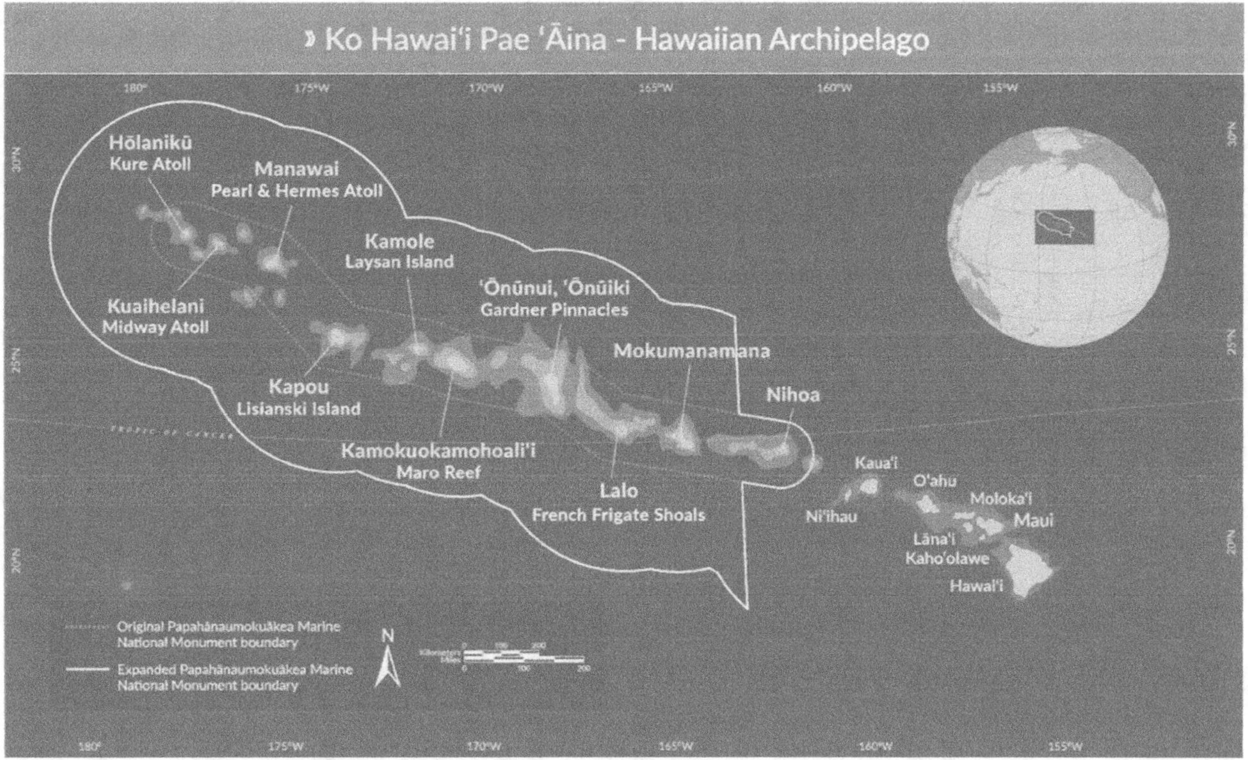

Figure 8.2 Map of Papahānaumokākea Marine National Monument
Source: US National Oceanic and Atmospheric Administration

claims to the islands, and to the ocean that surrounds them. Kānaka Maoli (Native Hawaiian) scientists have been appointed to leadership roles administering the reserve, outrigger canoe expeditions led by Kānaka Maoli are permitted to visit the islands, and funds have been made available to study the historical ritual significance of islands such as Nihoa and Mokumanamana. One Kānaka Maoli craftsman spent months on Midway (Kuaihelani) collecting 22,000 moulted albatross feathers to make kahili for the 'Iolani Palace.[18] The very name, Papahānaumokuākea, is a neologism coined to foster Indigenous pride in sites that were abandoned after the Christianisation of Hawaiian society in the early nineteenth century.[19] Anyone frustrated by the systemic archival silencing of Indigenous voices would feel hard-pressed not to celebrate such an endeavour.[20]

But Papahānaumokuākea is also one manifestation of what Bernard Oxman has called the 'territorial temptation', by which states exploit the provisions of UNCLOS to establish new powers over maritime spaces that were once regarded as the 'free seas'. For many environmentalists the new marine reserves represent something of a double-edged sword. Many do indeed help to protect threatened oceanic ecosystems from overfishing. But they are governed almost entirely unilaterally, and so hardly represent progress towards the kind of global protection that 'can best be realised through strong and effective international measures that states are obliged and empowered to enforce'.[21] And while the banning of longline fishing may have helped albatross numbers recover, the new park does nothing to defend low-lying atolls from rising sea levels or coral reefs from ocean acidification. Notwithstanding his enthusiasm for the exploits of Jacques Cousteau, it was George W. Bush who unilaterally withdrew the United States from the Kyoto Protocol. Arguably no single individual in history has done more to accelerate climate change.

[18] Mindy Pennybacker, 'Kahili Once More Grace Palace Throne', *Honolulu Star Advertiser* (30 November 2015): www.staradvertiser.com/2015/11/30/features/kahili-once-more-grace-palace-throne/.

[19] 'The Name': www.papahanaumokuakea.gov/new-about/name/. Last accessed 8 March 2018.

[20] Noelani Arista, 'Listening to Leoiki: Engaging Sources in Hawaiian History', *Biography* 32 no. 1 (2009): 66–73; Martin Dusinberre, 'Japan, Global History, and the Great Silence', *History Workshop Journal* 83 (Spring 2017): 130–50.

[21] Bernard H. Oxman, 'The Territorial Temptation: A Siren Song at Sea', *The American Journal of International Law* 100 no. 4 (2006): 844–5.

Finally, some marine reserves serve a decidedly ulterior purpose. In 2010 the Zoological Society of London announced a partnership with the UK government to establish a vast marine reserve in the Indian Ocean, centred around the British-administered Chagos Islands. The largest of these islands, Diego Garcia, was once the site of a coconut plantation worked by the descendants of enslaved and indentured labourers from Madagascar and East Africa.[22] But in the 1960s the Chagos Islanders were deported to make way for a US airbase that would become a pivotal part of the CIA's 'extraordinary rendition' network during the war on terror.[23] The base remains a key node in America's global 'pointillist empire' and anyone venturing into the marine reserve without permission will find themselves swiftly escorted from it by a US Navy patrol ship.[24]

But if the frigates protect the frigate birds the reverse is also true, for the US airbase on Diego Garcia has in recent years become subject to heightened legal and political scrutiny. Since the mid-2000s the deported Chagos Islanders have waged a remarkably effective campaign to be allowed to return home, mounting a series of legal challenges that culminated in favourable rulings by the UN General Assembly and the International Court of Justice.[25] The marine reserve was set up as an attempt to thwart this campaign, and thus protect the airbase. US State Department cables published by Wikileaks reveal UK diplomatic assurances that 'the British Indian Ocean Territory's former inhabitants would find it difficult, if not impossible, to pursue their claim for resettlement on the islands if the entire Chagos Archipelago were a marine reserve'.[26] It is hard to find a more chilling example of

[22] Laura Jeffery, '"We Are the True Guardians of the Environment": Human-Environment Relations and Debates About the Future of the Chagos Archipelago', *The Journal of the Royal Anthropological Institute* 19 no. 2 (2013): 302–3.

[23] Peter Sand, 'The Chagos Archipelago: Footprint of Empire, or World Heritage?' *Environmental Policy and Law* 40 no. 5 (2010): 322; Orla Guerin, 'Former Guantanamo Inmate Saad Iqbal Madni's "Agony"', *BBC News* (25 February 2011). Last accessed on 17 June 2020.

[24] Immerwahr, *How to Hide An Empire*, 18. For an account of an unauthorised visit to Diego Garcia see Simon Winchester, 'Diego Garcia', *Granta* 73 (22 December 2001): https://granta.com/diego-garcia/. Last accessed 21 April 2022.

[25] Owen Bowcott, 'UN Court Rejects UK's Claim of Sovereignty over Chagos Islands', *The Guardian* (25 February 2019).

[26] Wikileaks cable 09LONDON1156: 'HMG Floats Proposal for Marine Reserve' (15 May 2009).

nature conservation as a tactic for conserving sovereignty over territory.

The Chagos Islands are an extreme example, but access to most of the islands discussed in this book is now restricted to military and scientific personnel. Being neither, I did not hold out much hope of visiting one myself while researching this book. But in the summer of 2019, during a trip to the municipal archive in the Bonin Islands, I finally got my chance. Minamijima lies about twenty minutes by motorboat from Port Lloyd (J: Ōmura), the settlement founded by colonists from Hawai'i nearly two hundred years ago. With a surface area of a mere 28 hectares, it is small enough that the Tokyo metropolitan government was able to eradicate the island's rat population during its campaign to secure UNESCO heritage status for the Bonins. The island is now preserved as a haven for seabirds, one of the jewels in the crown of the Ogasawara National Park.[27]

Because of this, human settlement on Minamijima is now forbidden, but small groups of day-trippers are permitted to visit the island in the company of a tour guide. So it was that I – a historian by trade – was able to follow in the footsteps of the beachcombers, bird-hunters and scientists whose exploits I spent the best part of a decade studying. The guide shooed us along a narrow trail across sandy beaches and scrubby vegetation, while boobies and petrels wheeled and hooted overhead. There were no albatrosses to be seen, but an edge-tailed shearwater nested on the ground only a few feet from the trail. Vast privilege facilitated this encounter: an expensive education parlayed into an Ivy League job, which provided time and funds to travel round the world, roam far out to sea, in pursuit of my quarry. I stared at the bird, which seemed indifferent to my presence. 'Amazing,' I thought. 'I can put this in my book.'

[27] Nihon Shizen Hogo Kyōkai, *Ogasawara no takarabako Minamijima (17-nenkan no shizen kankyō monitaringu chōsa matome)* (Tokyo: Tōkyō-to Ogasawara Shichō, 2014), 2.

Appendix

Japanese Islands Abandoned, 1868–2013

Island Name	Municipality	Prefecture	Time Abandoned	Surface Area	Greatest Population	Notes
Funajima	Shimonoseki	Yamaguchi	1868–1912	0.08 km^2	Approx. 10	
Otōtojima	Ogasawara-mura (Bonin Islands)	Tokyo	1920–1926?	5.20 km^2	50~100	
Anijima	"	Tokyo	1920–1926?	7.87 km^2	2	
Anejima	"	Tokyo	1920–1926?	1.43 km^2	14	
Imōtojima	"	Tokyo	1920–1926?	1.22 km^2	33	
Meijima	"	Tokyo	1920–1926?	1.11 km^2	25	
Tsubajima	Iwagi-mura	Ehime	1926–1945	0.45 km^2	4 households	
Minami-torishima (Marcus Island)	Ogasawara-mura (Bonin Islands)	Tokyo	1933	1.10 km^2	70	Weather station and military facilities remain
Torishima	Hachijō-shichō	Tokyo	1939	4.79 km^2	125	Weather station in operation 1945–1965
Uotsuri-jima	Ishigaki-shi	Okinawa	1940?	3.82 km^2	248	Part of Senkaku (Ch: Diaoyu) island group
Mukojima	Ogasawara-mura (Bonin Islands)	Tokyo	1944	2.57 km^2	25	
Nakōdojima	"	Tokyo	1944	1.37 km^2	27	
Kita Iōtō	"	Tokyo	1944	5.57 km^2	220	

Yakabi-jima	Zamami-mura	Okinawa	1944–1945?	1.26 km^2	>2,000	
Sotopanari-jima	Taketomi-chō	Okinawa	1944–1945?	1.32 km^2	??	
Kami-mizushima	Kurashiki-shi	Okayama	1944–1945?	0.46 km^2	268	
Tsubura-jima	Fukue-shi	Nagasaki	1944–1945?	1.47 km^2	13 households	
Yomejima	Ogasawara-mura (Bonin Islands)	Tokyo	1944	0.81 km^2	20	
Iōtō (Iwo Jima)	”	Tokyo	1945	22.36 km^2	1,100	Military base remains
Ōkuno-jima	Takehara-shi	Hiroshima	1945	0.70 km^2	??	
Okidaitōjima (Rasa Island)	Daitō-mura	Okinawa	1945	1.15 km^2	>2,000	Used for bombing practice by US military
Hōgashima	Ushibuka-shi	Kumamoto	c.1955	0.26 km^2	4	
Mitsugo-jima	Ondō-cho	Hiroshima	1956	不明	65 households	
Ōshima	Mugi-chō	Tokushima	1958	1.76 km^2	23households	
Toshima	Ushibuka-shi	Kumamoto	1958	0.47 km^2	12	
Arashima	Wakamatsu-chō	Nagasaki	1959	0.15 km^2	12	
Iō-torishima	Gushikawa-son	Okinawa	1959	2.55 km^2	>600	
Hishima	Hagi-shi	Yamaguchi	1960	0.19 km^2	38	
Nuno-shima	Yuki-chō	Tokushima	1960	0.10 km^2	??	
Kashiwajima	Naoshima-chō	Kagawa	c.1960	0.48 km^2	<10	
Terashima	”	Kagawa	c.1960	0.23 km^2	>10	

(*cont.*)

Island Name	Municipality	Prefecture	Time Abandoned	Surface Area	Greatest Population	Notes
Kyōnojōrō-jima	”	Kagawa	c.1960	0.34 km^2	7	
Ōtsuku-jima	Hachiyo-shi	Kumamoto	1961	0.41 km^2	250	
Ōminase-jima	Tōwa-chō	Yamaguchi	1962	0.55 km^2	130	
Matajima	Kannonji-shi	Kagawa	1962	0.08 km^2	17	
Hyakkan-jima	Yuge-chō	Ehime	1962	0.03 km^2	5	
Ubushima	Kawaura-machi	Kumamoto	1962	1.94 km^2	3	
Maejima	Tokashiki-son	Okinawa	1962	1.60 km^2	300	
Uji-shima	Fukuyama-shi	Hiroshima	1963	0.52 km^2	41	
Ūjima	Ojika-chō	Nagasaki	1964	0.09 km^2	4	
Heichi-jima	Imabari-shi	Ehime	1965	0.25 km^2	??	
Yuri-jima	Nakashima-chō	Ehime	1965	0.45 km^2	75	
Ōitsukami-jima	Uwajima-shi	Ehime	1965	1.32 km^2	60	
Himeshima	Kishiku-chō	Nagasaki	1965	0.46 km^2	300	
Wadajima	Ōyano-machi	Kumamoto	1967	0.01 km^2	5	
Hachijō-kojima	Hachijō-machi	Tokyo	1969	3.08 km^2	513	
Usujima	Higashino-chō	Hiroshima	1969	0.46 km^2	46	
Yokoshima	Kurahashi-chō	Hiroshima	1970	0.59 km^2	130	
Kusumori-jima	Ryūgatake-machi	Kumamoto	1970	0.19 km^2	13	
Gaja-jima	Toshima-mura	Kagoshima	1970	4.07 km^2	82	

Gushikawa-shima	Izena-son	Okinawa	1970	0.42 km^2	24	
Kihē-jima	Naoshima-chō	Kagawa	c.1970	0.17 km^2	>100	
Sakumijima	Higashino-chō	Hiroshima	1971	0.31 km^2	25	
Funashima	”	Hiroshima	1971	0.30 km^2	50	
Tsungiri-jima	Mitsushima-chō	Nagasaki	1971	0.01 km^2	4	
Hashima	Hagi-shi	Yamaguchi	1972	0.27 km^2	115	
Ōshima	Hagi-shi	Yamaguchi	1972	0.22 km^2	103	
Yaburogi-jima	Ojika-chō	Nagasaki	1972	0.47 km^2	173	
Mishima	Ōmura-shi	Nagasaki	1972	2.34 km^2	66	
Kazurashima	Naru-chō	Nagasaki	1973	0.45 km^2	278	
Hashima	Takashima-chō	Nagasaki	1974	0.06 km^2	5,259	AKA Gunkanjima ('Battleship Island')
Takeshima	Goshoura-machi	Kumamoto	1974	0.04 km^2	22	
Ōshima	Ushibuka-shi	Kumamoto	1974	0.47 km^2	389	
Takashima	Masuda-shi	Shimane	1975	0.39 km^2	125	
Kamigurojima	Shimokamagari-chō	Hiroshima	1976	0.30 km^2	43	
Oreshima	Kamigotō-chō	Nagasaki	1976	0.32 km^2	230	
Shisakajima	Miyakubo-chō	Ehime	1977	1.26 km^2	>4,000	
Nagarasu-jima	Katsumoto-chō	Nagasaki	1977	0.27 km^2	3	
Sejima	Matsushima-chō	Kumamoto	1978	0.32 km^2	15	
Mage-shima	Nishinoomote-shi	Kagoshima	1980	8.40 km^2	383	

(*cont.*)

Island Name	Municipality	Prefecture	Time Abandoned	Surface Area	Greatest Population	Notes
Uchinapari-jima	Taketomi-chō	Okinawa	1982	2.10 km^2	>2,000	
Tatsushima	Tachibana-chō	Yamaguchi	1985	0.20 km^2	71	
Kojima	Akkeshi-chō	Hokkaido	1985	0.05 km^2	10 households	
Ōkurokami-shima	Okimi-chō	Hiroshima	1987	7.25 km^2	??	
Tsurujima	Hinase-chō	Okayama	1990	0.10 km^2	9	
Kamashima	Kurashiki-shi	Okayama	c.1990	0.40 km^2	76	
Ejima	Naoshima-chō	Kagawa	c.1995	0.28 km^2	30	
Ushigakubi-jima	”	Kagawa	c.1995	0.36 km^2	??	
Nozaki-jima	Ojika-chō	Nagasaki	2001	6.78 km^2	648	
Hitsushima	Hagi-shi	Yamaguchi	c.2009	0.81 km^2	118	
Toyoshima	Yuge-chō	Ehime	c.2010	0.40 km^2	180	
Hiki-jima	Imabari-shi	Ehime	c.2010	0.30 km^2	>70	
Shinjima	Sakurajima	Kagoshima	2013	0.13 km^2	250	

Source: Enosawa Yukihiro, *Ritō to hō: Izu Shotō Ogasawara Shotō kara kenpō mondai o kangaeru* (Tokyo: Hōritsu Bunkasha, 2018), 117.

Select Bibliography

Archival Collections

BPBM Bernice Pauahi Bishop Museum, Honolulu, Hawai'i
HSA Hawai'i State Archives, Honolulu, Hawai'i
MOFA Diplomatic Archives of the Ministry of Foreign Affairs, Tokyo
NAJ National Archives of Japan, Tokyo
NARA-II The US National Archives and Records Administration Archive No. II, College Park, Maryland
NIDS National Institute of Defense Studies, Tokyo
OPA Okinawa Prefectural Archives, Haebaru, Okinawa
OEC Ogasawara-mura Education Committee Archives, Chichijima
TMA Tokyo Metropolitan Archives, Tokyo
THBO Tokyo Metropolitan Government Hachijō Branch Office, Hachijōjima
UHM University of Hawai'i at Mānoa Library, Mānoa, Hawai'i
UKNA National Archives, Kew, United Kingdom

Periodicals

Asahi Shimbun. Accessed via Kikuzo II Visual: https://database.asahi.com/help/eng/about_e.html
Bulletin of the International Committee for Bird Preservation
Chūō Kōron
Harper's Bazar
Hawai Shimpō
The Hawaiian Gazette. Accessed via Library of Congress: https://chroniclingamerica.loc.gov/
The Hawaiian Star. Accessed via Library of Congress: https://chroniclingamerica.loc.gov/
ICBP Asian Section News
The Japan Times [AKA *Nippon Times* 1943–1956]. Accessed via https://info.japantimes.co.jp/archives/

Ka Nupepa Kuokoa. Accessed via Ulukau Hawaiian Electronic Library: https://ulukau.org/index.php?l=en
Keizai Zasshi Daiyamondo
The London & China Telegraph
Mainichi Shimbun. Accessed via Maisaku: https://mainichi.jp/contents/edu/maisaku/
Nature Scientific Reports. Accessed via www.nature.com/srep/
Nankai Taimusu
Nanryū Taimusu
Nihon Keizai Shimbun. Accessed via www.nikkei.com
Okinawa Taimusu
Ryukyū Shimpō
Sankei Shimbun. Accessed via www.sankei.com/
Smithsonian Magazine. Accessed via www.smithsonianmag.com/
South China Morning Post. Accessed via www.scmp.com
Yaeyama Bunka
Yomiuri Shimbun. Accessed via https://database.yomiuri.co.jp/about/en/

Books, Articles and Dissertations

Adelman, Jeremy and Stephen Aron. 'From Borderlands to Borders: Empires, Nation-States, and the Peoples in Between in North American History.' *American Historical Review* 104 no. 3 (June 1999): 814–41.

Aikawa Sadaharu and Fuse Rokurō. *Yoyogi Kōen*. Tokyo: Gōgakusha, 1981.

Akamine, Mamoru, Lina Terrell and Robert N. Huey. *The Ryukyu Kingdom: Cornerstone of East Asia*. Honolulu: University of Hawai'i Press, 2019.

Ambaras, David. *Japan's Imperial Underworlds: Intimate Encounters at the Borders of Empire*. Cambridge: Cambridge University Press, 2018.

Amrith, Sunil S. *Crossing the Bay of Bengal: The Furies of Nature and the Fortunes of Migrants*. Cambridge, MA: Harvard University Press, 2013.

Anderson, Valerie E.R. 'The Eurasian Problem in Nineteenth Century India', PhD diss., The School of Oriental and African Studies, 2011.

Ankei, Yuji. 'The Fieldwork Ethics and Positionality of Tsuneichi Miyamoto: A Pioneer of Border Island Studies in Japan.' *Eurasia Border Review* 4 no. 1 (2013): 19–42.

Aoki Eiji. *Yamashina Yoshimaro no shōgai*. Tokyo: Yamashina chōrui kenkyūjo, 1982.

Arch, Jakobina. *Bringing Whales Ashore: Oceans and the Environment of Early Modern Japan*. Seattle: University of Washington Press, 2018.

Asō Yawata. *Rinkō*. Tokyo: Maruzen, 1940.

Austin Jr, Oliver L. 'The Status of Steller's Albatross.' *Pacific Science* 3 (1949): 283–95.

Azuma, Eiichiro. *In Search of Our Frontier: Japanese America and Settler Colonialism in the Construction of Japan's Borderless Empire*. Oakland: University of California Press, 2019.

Banner, Stuart. 'Preparing to be Colonized: Land Tenure and Legal Strategy in Nineteenth-Century Hawaii.' *Law & Society Review* 39 no. 2 (2005): 273–314.

Barnhart, Michael A. *Japan Prepares for Total War: The Search for Economic Security, 1919-1941*. Ithaca: Cornell University Press, 2001.

Barrow, Mark. *A Passion for Birds: American Ornithology After Audubon*. Princeton: Princeton University Press, 2000.

Nature's Ghosts: Confronting Extinction from the Age of Jefferson to the Age of Ecology. Chicago: University of Chicago Press, 2009.

Benton, Lauren. *A Search for Sovereignty: Law and Geography in European Empires, 1400-1900*. Cambridge: Cambridge University Press, 2009.

Beckert, Sven, Ulbe Bosma, Mindi Schneider and Eric Vanhaute. 'Commodity Frontiers and the Transformation of the Global Countryside: A Research Agenda.' *Journal of Global History* 16 no. 3 (November 2021): 435–50.

BirdLife International. *Threatened Birds of Asia: The BirdLife International Red Data Book*. Cambridge: BirdLife International, 2001.

Bix, Herbert P. *Hirohito and the Making of Modern Japan*. New York: Harper Collins, 2000.

Boardman, Robert. *The International Politics of Bird Conservation: Biodiversity, Regionalism and Global Governance*. Cheltenham: Edward Elgar Publishing, 2006.

Botsman, Daniel V. *Punishment and Power in the Making of Modern Japan*. Princeton: Princeton University Press, 2007.

Brooke, Michael. *Albatrosses and Petrels Across the World*. New York: Oxford University Press, 2004.

Bryan, William Alanson. 'A Monograph of Marcus Island: An Account of Its Physical Features and Geology, With Descriptions of the Fauna and Flora.' *Occasional Papers of the Bernice Pauahi Bishop Museum* II no. 1 (1904): 77–139.

Bukh, Alexander. *These Islands Are Ours: The Social Construction of Territorial Disputes in Northeast Asia*. Stanford: Stanford University Press, 2020.

Burkman, Thomas W. *Japan and the League of Nations: Empire and World Order, 1914–1938*. Honolulu: University of Hawai'i Press, 2008.

Caldeira, Leah. *Royal Hawaiian Featherwork: Na Hulu Ali'i*. Honolulu: University of Hawai'i Press, 2015.

Cassel, Pär Kristoffer. *Grounds of Judgment: Extraterritoriality and Imperial Power in Nineteenth-Century China and Japan (Oxford Studies in International History)*. Oxford: Oxford University Press, 2012.

Chang, David. *The World and All the Things upon It: Native Hawaiian Geographies of Exploration*. Minneapolis: University of Minnesota Press, 2016.

Chapman, David. *The Bonin Islanders, 1830 to the Present: Narrating Japanese Nationality*. Lanham: Lexington Books, 2016.

Chemillier-Gendreau, Monique. *Sovereignty Over the Paracel and Spratley Islands*. Brussels: Groupe de Boeck, 2000.

Clark, Brett and John Bellamy Foster. 'Ecological Imperialism and the Global Metabolic Rift: Unequal Exchange and the Guano/Nitrates Trade.' *International Journal of Comparative Sociology* 50 (2009): 311–34.

Clulow, Adam. *The Company and the Shogun: The Dutch Encounter with Tokugawa Japan*. New York: Columbia University Press, 2014.

Courmont, Barthélémy. 'Territorial Disputes and Taiwan's Regional Diplomacy: The Case of the Senkaku/Diaoyu/Diaoyutai Islands.' *The Journal of Territorial and Maritime Studies* 1 no. 1 (2014): 113–34.

Cronon, William. 'The Trouble with Wilderness; or, Getting Back to the Wrong Nature', *Environmental History* 1 no. 1 (April 2019): 7–28.

Crosby, Alfred W. *Ecological Imperialism: The Biological Expansion of Europe, 900-1900*. Cambridge: Cambridge University Press, 2004.

Cushman, Gregory T. *Guano and the Opening of the Pacific World: A Global Ecological History*. Cambridge: Cambridge University Press, 2013.

Daniels, Roger. *The Politics of Prejudice: The Anti-Japanese Movement in California and the Struggle for Japanese Exclusion*. Oakland: University of California Press 1978.

David, Lance E., Robert E. Gallman and Karin Gleiter. *In Pursuit of Leviathan: Technology, Institutions, Productivity, and Profits in American Whaling, 1816–1906*. Chicago: University of Chicago Press, 1997.

Deguchi Tomohiro. 'Zekkai e no saidōnyū - ahōdori', in *Nihon no kishō chōrui o mamoru*, ed. Yamagiwa Satoshi. Tokyo: Tōkyō Daigaku Gakujutsu Shuppankai, 2009.

Demuth, Bathsheba. *Floating Coast: An Environmental History of the Bering Strait*. New York: W. W. Norton & Co., 2019.

'The Walrus and the Bureaucrat: Energy, Ecology, and Making the State in the Russian and American Arctic, 1870–1950.' *The American Historical Review* 124 no. 2 (April 2019): 483–510.

Doughty, Robin W. *Feather Fashions and Bird Preservation: A Study in Nature Protection*. Oakland: University of California Press, 1974.

Drayton, Richard. *Nature's Government: Science, Imperial Britain, and the 'Improvement' of the World*. New Haven: Yale University Press, 2000.

Drifte, Reinhard. 'The Japan-China Confrontation Over the Senkaku/ Diaoyu Islands - Between "Shelving" and "Dispute Escalation".' *Asia-Pacific Journal* 12 issue 30 no. 3 (28 July 2014).

Dudden, Alexis. 'Japan's Island Problem.' *Dissent* (Fall 2014).

Burnett, Christina Duffy. 'The Edges of Empire and the Limits of Sovereignty: American Guano Islands.' *American Quarterly* 57 no. 3 (September 2005): 779–803.

Dusinberre, Martin and Mariko Iijima. 'Transplantation and Imperial Practice in Japan's Pacific.' *Historische Anthropologie* 27 no. 3 (December 2019): 325–32.

Duus, Peter, Ramon H. Myers and Mark R. Peattie. *The Japanese Informal Empire in China, 1895–1937*. Princeton: Princeton University Press, 1989.

Elden, Stuart. *The Birth of Territory*. Chicago: University of Chicago Press, 2013.

Eldridge, Robert D. *Iwo Jima and the Bonin Islands in US–Japan Relations: American Strategy, Japanese Territory, and the Islanders In-Between*. Quantico: Marine Corps University Press, 2014.

Embree, John F. and Arthur C. Huston. 'Military Government in Saipan and Tinian: A Report on the Organization of Susupe and Churo, Together with Notes on the Attitudes of the People Involved.' *Applied Anthropology* 5 no. 1 (1946): 1–39.

Endō Masataka. *Koseki to kokuseki no kingendaishi: minzoku, kettō, Nihonjin*. Tokyo: Akashi Shoten, 2014.

Enosawa Yukihiro. *Ritō to hō: Izu Shotō Ogasawara Shotō kara kenpō mondai o kangaeru*. Tokyo: Hōritsu Bunkasha, 2018.

Esselstrom, Erik. *Crossing Empire's Edge: Foreign Ministry Police and Japanese Expansionism in Northeast Asia*. Honolulu: University of Hawai'i Press, 2008.

Fahmy, Ziad. 'Jurisdictional Borderlands: Extraterritoriality and "Legal Chameleons" in Precolonial Alexandria, 1840–1870.' *Comparative Studies in History and Society* 55 no. 2 (2013): 305–29.

Fedman, David. *Seeds of Control: Japan's Empire of Forestry in Colonial Korea*. Seattle: University of Washington Press, 2020.

Fesca, Max. *Nihon chisan ron*. Tokyo: Nōshōmusho Chisan Chōsasho, 1894.

Fitzmaurice, Andrew. *Sovereignty, Property, and Empire, 1500–2000*. Cambridge: Cambridge University Press, 2017.

Ford, Lisa. 'Law', *In Pacific Histories: Ocean, Land, People*, eds. David Armitage and Alison Bashford. London: Palgrave Macmillan, 2014.

Settler Sovereignty: Jurisdiction and Indigenous People in America and Australia, 1788-1836. Cambridge, MA: Harvard University Press, 2010.

Fujisawa Tōru. *Ahōdori*. Tokyo: Tōkō Shoin, 1967.

Gallagher, John and Ronald Robinson. 'The Imperialism of Free Trade.' *The Economic History Review New Series* 6 no. 1 (January 1953): 1–15.

Golovkin, A. N. 'The Effect of Colonial Seabirds on the Development of Phytoplankton.' *Oceanology* 7 (1967): 521–52.

Gonschor, Lorenz. *A Power in the World: The Hawaiian Kingdom in Oceania*. Honolulu: University of Hawai'i Press, 2019.

Grove, Richard H. *Green Imperialism: Colonial Expansion, Tropical Island Edens and the Origins of Environmentalism, 1600–1860*. Cambridge: Cambridge University Press, 1996.

Grzimek, Bernhard. *Grzimek's Animal Life Encyclopedia Vol. VII: Birds*. New York: van Nostrand Reinhold, 1975.

Guo, Nanyan. 'Environmental Culture and World Heritage in Pacific Japan: Saving the Ogasawara Islands.' *The Asia-Pacific Journal: Japan Focus* 7 issue 17 no. 3 (12 April 2009).

Hämäläinen, Pekka and Samuel Truett. 'On Borderlands.' *The Journal of American History* 98 no. 2 (2011): 361–38.

Hara, Kimie. 'Okinawa, Taiwan, and the Senkaku/Diaoyu Islands in US–Japan–China Relations.' *Asia-Pacific Journal: Japan Focus* 113 no. 28 (July 2015).

Hattori Tōru. 'Bakadori no hanashi.' *Dōbutsugaku Zasshi* 12 (15 October 1889): 405–11.

Havens, Thomas R. H. *Fire Across the Sea: The Vietnam War and Japan 1965–1975*. Princeton: Princeton University Press, 2014.

Parkscapes: Green Spaces in Modern Japan. Honolulu: University of Hawai'i Press, 2011.

Hayton, Bill. *The South China Sea: The Struggle for Power in Asia*. New Haven: Yale University Press, 2014.

Heise, Ursula. *Imagining Extinction: The Cultural Meanings of Endangered Species*. Chicago: University of Chicago Press, 2016.

Higuchi, Toshihiro. 'Japan as an Organic Empire: Commercial Fertilizers, Nitrogen Supply, and Japan's Core-Peripheral Relationship', in

Environment and Society in the Japanese Islands: From Prehistory to the Present, 139–157. Corvallis: University of Oregon Press, 2015.

Hirama, Yōichi. 'Japanese Naval Preparations for World War II.' *Naval War College Review* 44 no. 2 (1991): 63–81.

Hiraoka Akitoshi. *Ahōdori to 'teikoku' Nihon no kakudai: Nan'yō no shimajima e no shinshutsu kara shinryaku e*. Tokyo: Akashi Shoten, 2012.

Japanese Advance into the Pacific Ocean: The Albatross and the Great Bird Rush. Singapore: Springer, 2018.

'Meiji ki ni okeru Senkaku shotō e no Nihon-jin no shinshutsu to Koga Tatsushirō.' *Jinbun Chiri* 57 no. 5 (2005): 45–60.

Howell, David L. *Capitalism from Within: Economy, Society, and the State in a Japanese Fishery*. Oakland: University of California Press.

'Foreign Encounters and Informal Diplomacy in Early Modern Japan.' *The Journal of Japanese Studies* 40 no. 2 (Summer 2014): 295–327.

Geographies of Identity in Nineteenth-Century Japan. Oakland: University of California Press, 2005.

Howland, Douglas. *International Law and Japanese Sovereignty: The Emerging Global Order in the 19th Century*. London: Palgrave Macmillan, 2016.

Hutchinson, George. 'The Biogeochemistry of Vertebrate Excretion.' *Bulletin of American Natural History* 96 (1950): 1–554.

Immerwahr, Daniel. *How to Hide an Empire: A History of the Greater United States*. New York: Farrar, Straus and Giroux, 2019.

Inamura Hiromoto. 'Ogasawara-jima ni okeru shiseki oyobi enkaku.' *Rekishi Chiri* 48 (1926): 231–8.

Inoue, Masamichi S. *Okinawa and the US Military: Identity Making in the Age of Globalization, With a New Preface*. New York: Columbia University Press, 2017.

Iriye, Akira. *Pacific Estrangement: Japanese and American Expansion, 1897–1911*. Cambridge, MA: Harvard University Press, 1972.

Ishihara Shun. *'Guntō' no rekishi shakaigaku: Ogasawara Shotō, Iwōtō, Nihon, America, soshite Taiheiyō Sekai*. Tokyo: Kōbundō, 2013.

Iōtō: kokusaku ni honrō sareta 130 nen. Tokyo: Chūkō Shinsho, 2019.

Kindai Nihon to Ogasawara-tō: idōmin no shimajima to teikoku. Tokyo: Heibonsha, 2007.

Ishikawa Daisuke. 'Meiji-ki jinzō hiryō tokuyaku hanbai mo no seiritsu to tenkai: Ibaragi-ken Chiba-ken chiiki no jirei.' *Tochi seido shigaku* 173 (October 2001): 33–47.

Isono Naohide. *Tonosama seibutsugaku keifu*. Tokyo: Kagaku Asahi Henshūbu, 1991.

Izu Shotō Tōkyō Ikan 100-nen Shi Hensan Iinkai. *Izu Shotō Tōkyō ikan 100-nen shi*. Tokyo: Tōkyō-to Tōsho Chōson-kai, 1981.

Jacoby, Karl. *Crimes Against Nature: Squatters, Poachers, Thieves, and the Hidden History of American Conservation*. Oakland: University of California Press, 2001.

Jones, Ryan Tucker. *Empire of Extinction: Russians and the North Pacific's Strange Beasts of the Sea, 1741–1867*. Oxford: Oxford University Press, 2014.

'Running into Whales: The History of the North Pacific from Below the Waves.' *The American Historical Review* 118 no. 2 (April 2013): 349–77.

Kapur, Nick. *Japan at the Crossroads: Conflict and Compromise after Anpo*. Cambridge, MA: Harvard University Press, 2018.

Katsu Kaishū. *Kaigun rekishi* 4 no. 10. Tokyo: Kaigunshō, 1889.

Katsuragi Nami. 'Senkaku repōto – Tōkyō-to Senkaku Shotō yōjō chōsa' (6 September 2012).

Kawasaki Fusagorō. *Edo jidai no Hachijōjima: kotō ku no kyūmei*. Tokyo: Tōkyō-to, 1964.

Kayaoglu, Turan. *Legal Imperialism: Sovereignty and Extraterritoriality in Japan, the Ottoman Empire and China*. Cambridge: Cambridge University Press, 2014.

Kikiloi, Kekuewa. 'Rebirth of an Archipelago: Sustaining a Hawaiian Cultural Identity for People and Homeland.' *Hūlili* 6 (2010): 73–115.

Kikuchi Sakujirō. *Bakumatsu Ogasawaratō nikki*. Tokyo: Ryokuchisha, 1983.

Kirby, Peter Wynn. *Troubled Natures: Waste, Environment, Japan*. Honolulu: University of Hawai'i Press, 2010.

Kokusai Kōhō Kenkyūshitsu. *Minami-torishima jiken: Hōka Daigaku kokusai kōhō enshū hōkoku*. Tokyo: Tōkyō Teikoku Daigaku Kōnai Kokusai Kōhō Kenkyūshitsu, 1902.

Koskenniemi, Martti. *The Gentle Civilizer of Nations: The Rise and Fall of International Law, 1870–1960*. Cambridge: Cambridge University Press, 2001.

Kramer, Scott and Hanae Kurihara Kramer. 'The Other Isles of Aloha.' *The Journal of Hawaiian History* 47 (2013): 1–26.

Kublin, Hyman. 'The Ogasawara Venture (1861–1863).' *Harvard Journal of Asiatic Studies* 14 no. 1/2 (1951): 261–84.

Kumazawa Kikuo. 'Tsunetō Noritaka to Nihon no rinsan shigen.' *Hiryō Kagaku* 6 (1983): 27–68.

Kuykendall, Ralph S. *The Hawaiian Kingdom Vol. I, 1778-1854: Foundation and Transformation.* Honolulu: University of Hawai'i Press, 1938.

Lafeber, Walter. *The Clash: A History of US–Japan Relations.* New York: W. W. Norton & Co., 1997.

Latour, Bruno. *The Politics of Nature: How to Bring the Sciences into Democracy.* Cambridge, MA: Harvard University Press, 2004.

Lee, Ivy and Ming, Fang. 'Deconstructing Japan's Claim of Sovereignty Over the Diaoyu/Senkaku Islands.' *The Asia-Pacific Journal: Japan Focus* 10 no. 53 (30 December 2012): 1–49.

Loo, Tze May. *Heritage Politics: Shuri Castle and Okinawa's Incorporation into Modern Japan, 1879–2000.* Lanham: Lexington Books, 2014.

Lu, Sidney Xu. *The Making of Japanese Settler Colonialism: Malthusianism and Trans-Pacific Migration, 1868–1961.* Cambridge: Cambridge University Press, 2019.

Maier, Charles. 'Leviathan 2.0', In *A World Connecting, 1870–1945*, ed. Emily S. Rosenberg. Cambridge. MA: Harvard University Press, 2012.

Marcon, Federico. 'Satō Nobuhiro and the Political Economy of Natural History in Nineteenth-Century Japan.' *Journal of Japanese Studies* 34 no. 3 (December 2014): 265–87.

The Knowledge of Nature and the Nature of Knowledge in Early Modern Japan. Chicago: University of Chicago Press, 2015.

Matsuda, Matt K. *Pacific Worlds: A History of Seas, Peoples, and Cultures.* Cambridge: Cambridge University Press, 2012.

Matsuyama Shirō. *Yachō to tomo ni 80 nen: kiitari mitari tameshitari.* Tokyo: Bun'ichi Sōgō Shuppan, 1997.

McCormack, Gavan and Satoko Oka Norimatsu. *Resistant Islands: Okinawa Confronts Japan and the United States.* Lanham, MD: Rowman & Littlefield, 2018.

McKeown, Adam. 'Global Migration, 1846–1940.' *Journal of World History* 15 no. 2 (June 2004): 155–89.

'Mobility' , in *Pacific Histories: Ocean, Land, People*, ed. David Armitage and Alison Bashford. Basingstoke: Palgrave Macmillan, 2014.

McNeill, John. R. 'Of Rats and Men: A Synoptic Environmental History of the Island Pacific.' *Journal of World History* 5 no. 2 (1994): 299–349.

Melillo, Edward D. 'The First Green Revolution: Debt Peonage and the Making of the Nitrogen Fertilizer Trade, 1840–1930.' *American Historical Review* 117 (October 2012): 1028–60.

Miller, Ian Jared. *The Nature of the Beasts: Empire and Exhibition at the Tokyo Imperial Zoo.* Berkeley: University of California Press, 2011.

Miyamoto Tsuneichi. *Ritō ronshū.* Tokyo: Mizunowa Shuppan, 2009.

Miyamoto, Tsuneichi. *The Forgotten Japanese: Encounters with Rural Life and Folklore*. Berkeley: Stone Bridge, 2010.

Mochizuki Masahiko. 'Tamaoki Han'emon to Torishima kaitaku.' *Nantō Shigaku* 40 (October 1992): 40–59.

Moerman, Max D. 'Demonology and Eroticism: Islands of Women in the Japanese Buddhist Imagination.' *Japanese Journal of Religious Studies* 36 no. 2 (2009): 351–380.

Morgan, William Michael. *Pacific Gibraltar: US–Japanese Rivalry over the Annexation of Hawai'i, 1885–1898*. Annapolis: Naval Institute Press, 2011.

Morita Yoshio. *Rasa-tō shubitai ki: gyokusai o kakugo shita heishitachi no ningen dorama*. Tokyo: Kōjinsha NF Bunko, 1995.

Morris-Suzuki, Tessa. *Borderline Japan: Frontier Controls, Foreigners and the Nation in the Post-War Era*. Cambridge: Cambridge University Press, 2010.

On the Frontiers of History: Rethinking East Asian Borders. Canberra: ANU Press, 2020.

Mulder, Nicholas. 'The Economic Weapon: Interwar Internationalism and the Rise of Sanctions, 1914–1945', PhD diss., Columbia University, 2019.

Murai Osamu. *Nantō ideorogii no hassei: Yanagita Kunio to shokuminshugi*. Fukutake Shoten, 1992.

Nakayama Yoshitaka. *Chūgoku ga mimi o fusagu Senkaku Shotō no futsugōna shinjitsu: Ishigaki shichō ga tsutsuguru Nihon gaikō no aru beki sugata*. Tokyo: Wani Bukkusu PLUS Shinsho, 2012.

Neu, Charles E. *An Uncertain Friendship: Theodore Roosevelt and Japan, 1960–1909*. Cambridge, MA: Harvard University Press, 1967.

Ngo, Ti. 'Mapping Economic Development: The South Seas Government and Sugar Production in Japan's South Pacific Mandate, 1919–1941.' *Cross-Currents: East Asian History and Culture Review* 2 (March 2012): 1–23.

Nihon Kasei Hiryō Kyōkai. *Rinsan hiryō kōgyō no ayumi*. Tokyo: Nihon Kasei Hiryō Kyōkai, 1972.

Nishimura Kazuyuki. 'Taiwan higashi kaigan ni okeru gyorō gijutsu no bunka shigenka.' *Higashi Ajia Kindaishi* 17 no. 3 (2014): 38–55.

Obana Sakusuke. *Ogasawara Yōroku*. Tokyo: Yuūnsha, 2014.

Obermiller, David Tobaru. 'The US Military Occupation of Okinawa: Politicizing and Contesting Okinawan Identity, 1945-1955', PhD diss., University of Iowa, May 2006.

Ogasawara, Ko and Ikehara Sadao. 'Ecological and Behavioral Observations of Okinawa Woodpecker *Sapheopipo noguchii*, with Notes on Conservation.' *Yamashina Chōrui Kenkyū* 9 no. 2 (1977): 13–28.

Ogawa Tadashi. *Sengo Beikoku no Okinawa bunka senryaku: Ryūkyū Daigaku to Michigan misshon*. Tokyo: Iwanami Shoten, 2012.

Ohara Tomoko. 'Sengo Okinawa shakai to Nanyō Guntō Hikiagesha - hikiagesha dantai katsudō ni chūmoku shite.' *Imin Kenkyū* 6 (March 2010): 23–44.

Ōkawa Kazushi. *Nōrinsuisanshō chōki keizai tōkei: Nōringyō* vol. 9. Tokyo: Tōyō Keizai Shinpōsha, 1965.

Okura Unosuke. *Bōfū no shima*. 1940.

Okuyama Ikuko. 'Hachijōjima ni okeru jinkō ryūshutsu katei to sono tokushitsu: omo ni Meiji shoki kara Dai-Ni Sekai Taisen made.' *Chigaku Zasshi* 95 no. 1 (1986): 46–61.

Oshikawa Shunrō. *Kaitei Gunkan: Kaitō Bōken Kitan*. Tokyo: Bunbudo, 1900.

Osterhammel, Jürgen. *The Transformation of the World: A Global History of the Nineteenth Century*. Princeton: Princeton University Press, 2014.

Ostrom, Elinor. *Governing the Commons: The Evolution of Institutions for Collective Action*. Cambridge: Cambridge University Press, 1990.

Oxman, Bernard H. 'The Territorial Temptation: A Siren Song at Sea.' *The American Journal of International Law* 100 no. 4 (2006): 830–51.

Ōzawa Junji. *Guano hiryō: Minami Torishima san chōfun rinkō*. Tokyo: Zenkoku Hiryō Toritsugisho Kabushiki Gaisha, 1904.

Pan, Zhongqi. 'Sino-Japanese Dispute Over the Diaoyu/Senkaku Islands: The Pending Controversy from the Chinese Perspective.' *Journal of Chinese Political Science* 12 no. 1 (2007): 71–92.

Peattie, Mark R. *Nan'yō: The Rise and Fall of the Japanese in Micronesia, 1885–1945*. Honolulu: University of Hawai'i Press, 1992.

Pedersen, Susan. *The Guardians: The League of Nations and the Crisis of Empire*. Oxford: Oxford University Press, 2015.

Perez, Louis G. *Japan Comes of Age: Mutsu Munemitsu and the Revision of the Unequal Treaties*. Vancouver: Fairleigh Dickinson University Press, 1999.

Petersen, Glen. 'Indigenous Island Empires: Yap and Tonga Considered.' *The Journal of Pacific History* 35 no. 1 (June 2000): 5–27.

Petty, Bruce M. *Saipan: Oral Histories of the Pacific War*. Jefferson: McFarland & Co., 2009.

Press, Steven. *Rogue Empires: Contracts and Conmen in Europe's Scramble for Africa*. Cambridge, MA: Harvard University Press, 2017.

Rasa Kōgyō Kabushiki Gaisha Shashi Henshūshitsu. *Rasa Kōgyō 80-nen shi*. Tokyo: Rasa Kōgyō Kabushiki Gaisha, 1993.

Ravina, Mark. *To Stand with the Nations of the World: Japan's Meiji Restoration in World History*. New York: Oxford University Press, 2017.

Reeves, Randall R. 'The Origins and Character of "Aboriginal Subsistence" Whaling: A Global Review.' *Mammal Review* 32 no. 2 (June 2002): 71–106.

Rinyachō. *Chōjū gyōsei no ayumi*. Tokyo: Rinya Kōsai-kai, 1979.

Roberts, Luke S. *Performing the Great Peace: Political Space and Open Secrets in Tokugawa Japan*. Honolulu: University of Hawai'i Press, 2015.

Rosenthal, Gregory T. *Beyond Hawai'i: Native Labor in the Pacific World*. Oakland: University of California Press, 2018.

'Life and Labor in a Seabird Colony: Hawaiian Guano Workers, 1857–70.' *Environmental History* 17 (2012): 744–82.

Rüegg, Jonas. 'Business, State and Environment in the Making of Japan's Kuroshio Frontier.' In *Oceanic Japan*, ed. Stefan Hubner, Nadin Heé, Ian Jared Miller and William Tsutsui. Honolulu: University of Hawai'i Press, forthcoming.

'Mapping the Forgotten Colony: The Ogasawara Islands and the Tokugawa Pivot to the Pacific.' *Crosscurrents: East Asian History and Culture Review* 23 (June 2017).

Ryūkyū Seifu Bunkazai Hogo Iinkai. *Bunkazai yōran*. Naha Keitai Hyōronsha, 1965.

Okinawa no bunkazai. Naha: Keitai Hyōronsha, 1967.

Sakagami, Shoichi F. 'An Ecological Perspective of Marcus Island, With Special Reference to Land Animals.' *Pacific Science* 15 no. 1 (1961): 82–104.

Sakano Tōru. 'Kyū Shakai Rengō no kyōdō chōsa to "kokudo"'. *Nihon Daigaku Keizai Gakubu Sangyō Keiei Kenkyūjō Shohō* 263 (December 2011).

Samuels, Marwyn. *Contest for the South China Sea*. London: Routledge, 2013.

Sand, Peter. 'The Chagos Archipelago: Footprint of Empire, or World Heritage?' *Environmental Policy and Law* 40 no. 5 (2010): 232–42.

Saranillio, Dean Itsuji. 'Why Asian Settler Colonialism Matters: A Thought Piece on Critiques, Debates, and Indigenous Difference.' *Settler Colonial Studies* 3 no. 3–4 (2013): 280–94.

Sarantakes, Nicholas Evan. *Keystone: The American Occupation of Okinawa and US–Japanese Relations*. College Station: Texas A&M University Press, 2001.

Sasges, Gerard. 'Absent Maps, Marine Science, and the Reimagination of the South China Sea, 1922–1939.' *The Journal of Asian Studies* 75 no. 1 (February 2016): 157–80.

Sawada, Mitziko. 'Culprits and Gentlemen: Meiji Japan's Restrictions of Emigrants to the United States, 1891-1909'. *Pacific Historical Review* 60 no. 3 (August 1991): 339–59.
Schencking, J. Charles. *Making Waves: Politics, Propaganda, and the Emergence of the Imperial Japanese Navy, 1868-1922*. Stanford: Stanford University Press, 2005.
'The Imperial Japanese Navy and the Constructed Consciousness of a South Seas Destiny, 1872-1921'. *Modern Asian Studies* 33 no. 4 (1999): 769–96.
Scott, James C. *The Art of Not Being Governed: An Anarchist History of Upland Southeast Asia*. New Haven: Yale University Press, 2009.
Weapons of the Weak: Everyday Forms of Peasant Resistance. New Haven: Yale University Press, 1987.
Schlesinger, Jonathan. *A World Trimmed with Fur: Wild Things, Pristine Places, and the Natural Fringes of Qing*. Stanford: Stanford University Press, 2017.
Seifman, Travis. 'Performing "Lūchū": Identity Performance and Foreign Relations in Early Modern Japan.' PhD diss., University of California Santa Barbara, 2019.
Senanami Chōsen. 'Gunchō zasso.' *Yaeyama Bunka* 11 (May–June 1947).
Senkaku Shotō Bunken Shiryō Kenkyū Hensankai. *Senkaku kenkyū: Takara gakujutsu chōsadan shiryōshū*. Naha: Dētamu Rekishiosu, 2007.
Setoguchi Akihisa. 'Yachō o meguru dōbutsu-kan.' In *Nihon no dōbutsu-kan: hito to dōbutsu no kankeishi*, ed. Ishida Osamu, Hamano Sayoko, Hanazono Makoto and Setoguchi Akihisa. Tokyo: Tōkyō Daigaku Shuppankai, 2013.
Shiga Shigetaka. *Nan'yō jijō*. Tokyo: Maruzen, 1887.
Shiode Hiroyuki. *Ekkyosha no Seijishi: AjiaTaiheiyō ni okeru Nihonjin no imin to shokumin*. Nagoya: Nagoya Daigaku Shuppankai, 2015.
Shirane, Haruo. *Japan and the Culture of the Four Seasons: Nature, Literature, and the Arts*. New York: Columbia University Press, 2011.
Shirane, Seiji. *Imperial Gateway: Colonial Taiwan and Japan's Expansion in South China and Southeast Asia, 1895-1945*. Ithaca: Cornell University Press, 2022.
'Mediated Empire: Colonial Taiwan in Japan's Imperial Expansion in South China and Southeast Asia, 1895-1945.' PhD diss. Princeton University, 2014.
Skabelund, Aaron. *Empire of Dogs: Canines, Japan, and the Making of the Modern Imperial World*. Ithaca: Cornell University Press, 2011.
Skaggs, Jimmy M. *The Great Guano Rush: Entrepreneurs and American Overseas Expansion*. London: Palgrave Macmillan, 1994.

Smith, Bernard. *Imagining the Pacific: In the Wake of the Cook Voyages.* New Haven: Yale University Press, 1992.

Smith, Tim D., Randall R. Reeves, Elizabeth A. Josephson and Judith N. Lund. 'Spatial and Seasonal Distributions of American Whaling and Whales in the Age of Sail.' *PloS One* 7 no. 4 (April 2012): 1–25.

Smits, Gregory. *Maritime Ryukyu, 1050–1650.* Honolulu: University of Hawai'i Press, 2019.

Stein, Sarah Abrevaya. *Plumes: Ostrich Feathers, Jews, and a Lost World of Global Commerce.* New Haven: Yale University Press, 2008.

Stoll, Steven. *Larding the Lean Earth: Soil and Society in 19th-Century America.* New York: Farrar, Straus and Giroux, 2003.

Suganuma, Unryu. *Sovereign Rights and Territorial Space in Sino-Japanese Relations: Irredentism and the Diaoyu/Senkaku Islands.* Honolulu: University of Hawai'i Press, 2000.

Sugawara Hiroshi and Kakizawa Ryōzō. *Zusetsu chōmei no yurai jiten.* Tokyo: Kashiwa Shobō, 2005.

Sugiyama-Lebra, Takie. *Above the Clouds: Status Culture of the Modern Japanese Nobility.* Berkeley: University of California Press, 1995.

Takaesu Masaya. *Kindai Nihon no chihō tōchi to tōsho.* Yumani Shobō, 2009.

Takara Tetsuo 'Jinmin sensen no hata.' *Yaeyama Bunka* 1 (July 1946)

'Ryūkyū Nakanokamijima no kaichō.' *Yamashina Chōrui Kenkyū Hōkoku* 6 no. 1–2 (1970): 188–94.

Yaeyama kaihatsu shiryō. n.d.

Tanaka Hiroyuki. *Bakumatsu no Ogasawara: Ōbei no hogeisen de sakaeta midori no shima.* Tokyo: Chūō Kōron, 1997.

Tanigawa Ken'ichi. *Hyōryū to hyōchaku sōsakuin.* Tokyo: Shōgakukan, 1993.

Teaiwa, Katerina Martina. *Consuming Phosphate Island: Stories of People and Phosphate from Banaba.* Bloomington: Indiana University Press, 2014.

Thomas, Julia Adeney. *Reconfiguring Modernity: Concepts of Nature in Japanese Political Ideology.* Oakland: University of California Press, 2001.

Thomas, Nicholas. *Islanders: The Pacific in the Age of Empire.* New Haven: Yale University Press, 2010.

Thrush, Coll. *Indigenous London: Native Travelers at the Heart of Empire.* New Haven: Yale University Press, 2016.

Tickell, W. Lance N. *Albatrosses.* Mountfield: Pica Press, 2000.

Tierney, Robert. *Tropics of Savagery: The Culture of Japanese Empire in Comparative Frame.* Oakland: University of California Press, 2010.

Togawa Yukio. *Iriomote yamaneko.* Tokyo: Jiyū Kokuminsha, 1972.

Torishima Kishōchō Kurabu 'Torishima' Henshū Iinkai. *Torishima*. Tokyo: Tōe, 1967.
Totman, Conrad. *Japan's Imperial Forest: Goryōrin, 1889–1946: With a Supporting Study of the Kan/Min Division of Woodland in Early Meiji Japan, 1871–76*. Honolulu: University of Hawai'i Press, 2007.
Tōyō Keizai Shimpōsha. *Nihon bōeki yōran*. Tokyo: Tōyō Keizai Shimpōsha, 1975.
Tsujihara Makihiko and Imaizumi Yumiko. *Nan'Yōcho Kohō*. Tokyo: Yumanishobō, 2009.
Tsunetō Noritaka. *Jitsuyō hiryō no kōwa*. Tokyo: Chūgai Hiryō Yōhōsha, 1907.
'Meiji no Pedorojisuto.' *Pedorojisuto* 35: 1 1991.
Minami Nihon no fugen. Tokyo: Hakubunkan, 1910.
Nihon dojō ron. Tokyo: Seibidō, 1904.
Rinpi chōsa hōbun dai 1-gō: hinata guni rinkō. Tokyo: Nōshōmusho Chishitsu Chōsasho, 1896.
Tsunetō, Noritaka. *Discovery and Exploitation of Rasa Island*. Tokyo: Rasa Rinkō Kabushiki Gaisha, 1920.
Tsushima Hideko. 'Kita Mariana Shotō kara no hikiagesha: Hachijōjima tōmin no imin jirei o chūshin toshite.' *Shirayama Jinruigaku* 11 (March 2008): 147–66.
Tsutsui, William. 'The Pelagic Empire: Reconsidering Japanese Expansion.' In *Japan at Nature's Edge: The Environmental Context of a Global Power*, eds. Ian Jared Miller, Julia Adeney Thomas and Brett L. Walker. Honolulu: University of Hawai Press, 2013.
Uchida, Jun. 'From Island Nation to Oceanic Empire: A Vision of Japanese Expansion from the Periphery.' *Journal of Japanese Studies* 42 no. 1 (Winter 2016): 57–90.
Uchida Seinosuke. *Tennen kinenbutsu: chōrui-hen*. Tokyo: Sōgensha, 1960.
Uehara Keiji. *Zōen taikei*. Tokyo: Kashima Shoten, 1974.
Umō Kenkyūkai, ed. *Umō to negu no hanashi: sono rekishi to bunka*. Tokyo: Nihon Keizai Hyōronsha, 1993.
Unger, Tom. *Max Schlemmer, Hawaii's King of Laysan Island*. Bloomington: iUniverse, 2004.
Walker, Brett L. *The Conquest of Ainu Lands: Ecology and Culture in Japanese Expansion, 1590-1800*. Oakland: University of California Press, 2006.
The Lost Wolves of Japan. Seattle: University of Washington Press, 2005.
Walther, Michael and Julian P. Hume. *Extinct Birds of Hawaii*. Honolulu: Mutual Publishing, 2016.
Weil, Kari. *Thinking Animals: Why Animal Studies Now*. New York: Columbia University Press, 2012.

Wigen, Kären. 'Discovering the Japanese Alps: Meiji Mountaineering and the Quest for Geographical Enlightenment.' *The Journal of Japanese Studies* 31 no. 1 (2005): 1–26.

Wilson, Noell. *Defensive Positions: The Politics of Maritime Security in Tokugawa Japan*. Cambridge, MA: Harvard University Press, 2015.

'Whaling at the Margins: Drift Whales, Ainu Laborers, and the Japanese State on the Nineteenth-Century Okhotsk Coast.' In *New Histories of Pacific Whaling*, ed. Ryan Tucker Jones and Angela Wanhalla. Munich: Rachel Carson Center for Environmental History, 2019.

Winichakul, Thongchai. *Siam Mapped: A History of the Geo-Body of a Nation*. Honolulu: University of Hawai'i Press, 1997.

Xu Longsheng, 'Zhong-Ri liang guo weirao Xisha Qundao lin kuang kaifa de hezuo, jingzheng yu jiufen (1917–1930).' *Shi Lin* 5 (2017).

Yamada Yoshihiko. *Senkaku Shotō no kanri mondai ni kan suru hōkoku*. Tokyo: Tōkyō-to, 31 March 2013.

Yamamoto, Takahiro. '*Balance of Favour: The Emergence of Territorial Boundaries Around Japan, 1861-1875*', PhD diss., The London School of Economics, 2015.

Yamashina Yoshimaro. 'Chōrui hyōshiki shiken hōkoku dai 2-kai (Shōwa 37 nendo).' *Yamashina Chōrui Kenkyū* 3 no. 5 (December 1963).

Nihon no chorui to sono seitai. Tokyo: Azusa Shobo, 1934.

Tori no heru kuni fueru kuni: Ōbei tori angya. Tokyo: Nihon Chōrui Hogo Renmei, 1967.

Yokota Kisaburō. 'Mujintō sensen-ron.' *Chūō Kōron* (September 1933): 84–93.

Yokota Masatsugu and Yokohata Yasushi. 'Senkaku Shotō Uotsurishima no yaseika yagi mondai ni tsuite.' *Wildlife Conservation Japan* 5 no. 1–2 (2000): 1–12.

Yokohata, Yasushi, Yukio Ikeda, Masatsugu Yokota and Hideharu Ishizaki. 'Effects of Introduced Goats on the Ecosystem of Uotsuri-Jima in the Senkaku Islands, Japan, as Assessed By Remote-Sensing.techniques.' *Biosphere Conservation* 5 no. 1 (2003): 29–46.

Young, Louise. *Japan's Total Empire: Manchuria and the Culture of Wartime Imperialism* Oakland: University of California Press, 1999.

Index

For EU product safety concerns, contact us at Calle de José Abascal, 56–1°,
28003 Madrid, Spain or eugpsr@cambridge.org.

www.ingramcontent.com/pod-product-compliance
Lightning Source LLC
LaVergne TN
LVHW010713061126
851857LV00014B/1170
9781108747462